Diaspora in Development

Ajay Dubey · Sushmita Rajwar
Editors

Diaspora in Development

The Comparative Indian Experience

palgrave
macmillan

Editors
Ajay Dubey
School of International Studies
Jawaharlal Nehru University
New Delhi, Delhi, India

Sushmita Rajwar
School of International Studies
Jawaharlal Nehru University
New Delhi, Delhi, India

ISBN 978-981-96-6746-8 ISBN 978-981-96-6747-5 (eBook)
https://doi.org/10.1007/978-981-96-6747-5

This Palgrave Macmillan imprint is published by the registered company Springer Nature Singapore Pte Ltd.
The registered company address is: 152 Beach Road, #21-01/04 Gateway East, Singapore 189721, Singapore

Contents

Notes on Contributors

Aditya Anshu is an Assistant Professor of International Relations at Abu Dhabi University, Abu Dhabi. Dr. Anshu has completed M.Phil. and Ph.D. degree in International Relations from Jawaharlal Nehru University, and Master in Political Science. Dr. Anshu's research interests include State Politics, Foreign Policy, Diplomacy and Diaspora. Dr. Anshu is a member of several professional bodies, including the Organization for Diaspora Initiative, and the Migration and Diaspora Institute, New Delhi. Besides, he has also been appointed an Associate Editor of the *Journal of Critical South Asian Studies* in London. He recently edited the book, *Diaspora in the MENA Region* published by the University of Johannesburg Press and is working on a volume for Palgrave Macmillan titled *Interconnected Horizons: Navigating Globalization, Diaspora, Foreign Policy, and Sustainability*.

Prof. Chandrashekhar Bhat began his teaching career in 1973 at the University of Mysore, soon after the completion of his doctoral studies at the Department of Sociology, Delhi School of Economics, Delhi University. He established the Department of Sociology and Anthropology at the Central University of Hyderabad in 1979. Prof. Bhat's pioneering research and expertise in sociology of migration and diaspora studies were mentored particularly by Professor M.S.A. Rao at the Department of Sociology, Delhi University and Prof. Adrian C. Mayer under a Post-Doctoral program at School of Oriental and African Studies, London University, U.K. Prof. Bhat has introduced for the first time in India, a course on Overseas Indians as early as 1985 at the Department of Sociology, University of Hyderabad.

Ajay Dubey Prof. Ajay Kumar Dubey is currently Professor at School of International Studies, Jawaharlal Nehru University. He has held various academic positions in the past that includes Rector, JNU, New Delhi;); Director, Jawaharlal Nehru Institute of Advanced Study (JNIAS), JNU; Founding Chairperson, Special Centre for National Security Studies (SCNSS), JNU; Director, Energy Studies Programme, SIS, JNU; and Chairperson, Centre for African Studies, SIS, JNU.

Prof. Dubey has published several books related to international studies including on Diaspora issues- Indian Diaspora Contributions to their New Home (MD Publication, 2011), Indian Diaspora in Africa: A Comparative Perspective (MD Publication, 2009) and Indian Diaspora: Global Identity (Kalinga Publication, 2003).

He is also the book series editor of, "Africa's Global Engagement", by Palgrave-Macmillan; Editor-in-Chief of Diaspora Studies, a quarterly refereed journal published by Brill, Netherlands. Currently, he is also President of Organisation for Diaspora Initiative (ODI) which is focused on academic research and publication on Global Diasporas.

Prof. Dubey was also the Principal Investigator of Major Research Project Team under UGC project on '*Diaspora in Development*: Comparative Assessment of Harnessing Globalisation', 2014–2017 and Principal Investigator of UGC Major Project on 'India and Indian Diaspora in Africa' at School of International Studies, JNU, 2010–13.

Rafalimanana Jeannie is currently working as Chargé d'Affaires at the Embassy of Madagascar in Algiers. Previously she has also worked as a Diplomatic Agent at Ambassador of Madagascar where she was the Director of Diaspora department. She was also the President of HAVATSA–UPEM which is an organization of Malagasy writers and poets. Ms. Jeannie has a strong professional background as she has graduated from the International Institute of Humanitarian Law, Sanremo, Italy. She completed her diploma in Migration Management from the United Nations University. She is an experienced specialist with a demonstrated history of working in the government administration industry. Skilled in Nonprofit Organizations, Analytical Skills, Public Speaking, Creative Writing and International Relations.

Jeffery D. Long is a Professor of Religion and Asian Studies at Elizabethtown College, author of several books, and a 2018 recipient of the Dharma Seva Award for his work to dispel stereotypes about Hinduism. He was long graduated with a B.A. from the University of Notre Dame in 1991. During his time at Notre Dame, he converted from Catholicism to Hinduism, as later informed by his work. He received his M.A. and Ph.D. at the University of Chicago Divinity School in 1993 and 2000, respectively. His Ph.D. thesis was titled Plurality and Relativity: Whitehead, Jainism and the Reconstruction of Religious Pluralism. He began work at Elizabethtown College in 2000. Prof. Long has authored four books on Hinduism and Jainism. He was a consultant to the Hindu American Foundation during its lawsuit against the California Board of Education.

Manoranjan Mohanty joined Development Studies Program, University of South Pacific, Fiji in 1999. Previously, he was engaged in research and postgraduate teaching at Political Geography Division, School of International Studies, Jawaharlal Nehru University, New Delhi. Having both social science and natural science background, Mohanty has a wide range of specialism and areas of interest. He has extensive experience in research, postgraduate teaching and consultancy. He received the 2015 Vice Chancellor Award for Excellence in Teaching at USP. He maintains a sound publication profile and he has published seven books/monographs and many research papers to his credit. He has a keen interest towards development theories and policies. In his research and teaching, he focuses on the processes, linkages and patterns of change and development covering many aspects—social, economic, political and environmental from an interdisciplinary perspective and at local, national, regional and international levels. His research interests focus primarily on the developing world and in particular, the Asia-Pacific region. Currently, his research covers the Pacific Island countries especially Fiji, and Oceania as a whole.

Rashmi Mourya has completed M.Phil. and Ph.D. in African studies from Jawaharlal Nehru University, New Delhi. She has a Master's degree in Political Science. Currently, she is working as an Assistant Professor of Political Science at a College affiliated to Shaheed Nandkumar Patel University Raigarh, Chhattisgarh. She is also associated with the African Studies Association of India (ASA India). Her research interests include African studies, Diaspora Studies, Ethiopian Diaspora, Strategic issues and International Relations.

Sushmita Rajwar Dr. Sushmita Rajwar is currently working as Associate Professor, at School of International Studies. She has formerly worked as Associate Professor, Department of Politics & Governance, Central University of Kashmir and as Assistant Professor at Department of Political Science, Maharaja Agrasen College, University of Delhi. She has completed her Ph.D. from School of International Studies, Jawaharlal Nehru University, New Delhi. Her research interest beside other areas include India Africa relations, China in Africa and Indian Diaspora in Africa. She has been an active member of African Studies Association of India (ASA), an international network of higher academics which publishes international journal Diaspora Studies (Brill Publication, Netherlands) She has published several papers and presented papers in many conferences and workshops in India and abroad on topics related to Chinese Engagement in African countries and its comparison with Indian engagement.

Jean Regis Ramsamy is a journalist from Réunion islands who has written extensively about the Indians in Réunion. His articles have focused on the various needs of the Indian diaspora in Réunion and he has been able to provide some very useful insights into the policy making of the Indian government towards its diaspora in a small country like Réunion. His writings brought into light the conditions of the Indian diaspora in the Francophone regions of Africa as well. He also received the prestigious Pravasi Bharatiya Samman Award in 2021 for his contribution in the field of Media and community service.

Satwinder Rehal is an Associate Professor, School of Arts & Sciences at Philippine Women's University (PWU). He has also worked as a Senior lecturer at the University of the Philippines Open University. He specializes in international relations studies, particularly in International Relations Theory and cultural diplomacy. He holds a double Master's degree from the University of Queensland, Australia. He has completed his Doctorate in Sociology at the University of the Philippines, Diliman. Originally from Kenya, he received recognition for Best Paper Award during the 1st University of Malaysia, Sabah International Sports Science Conference on his paper entitled: "*Azkals Fever Phenomenon: Setting An Agenda on a Discourse in The Sociology of Soccer in the Philippines*" and was the winner of the 17th OSSREA Science Research Grant for Young Scholars in 2003 for a study entitled "*Mining-induced Resettlement Effects and Impoverished Risks: A case study of the Titanium Minerals*

Sands Project in Kwale, Kenya." He recently had a paper accepted and presented at the 2nd International Conference on ASEAN Studies held at Chulalongkorn University, Bangkok, Thailand in August, 2015 entitled "*Enhancing Cultural Diplomacy in the ASEAN Socio-Cultural Strategy: case of The Philippines' Bayanihan Dance Company and a proposed 'Rizal Institute' (Instituto Rizalista).*"

Santosh Kumar Singh is an Associate Professor at the Department of Social and Political Studies, Manav Rachana International Institute of Research and Studies. He has done an M.Phil. and Ph.D. from the School of International Studies, Jawaharlal Nehru University, New Delhi, India. He has undertaken research on governance, democracy, civil society, foreign policy and security issues of the countries in Africa. He has done intensive research on civil society and foreign policy issues in Sub-Saharan Africa with special focus on South Africa. His areas of research interest include India-Africa Relations, Indian Diaspora in Africa, governance, democracy, civil society and Conflict. He has presented research papers in various national and international seminars in India and abroad. He has contributed several articles in journals, magazines and chapters in edited books.

Rodolfo García Zamora is presently Director of the Academic Unit in Development Studies at the Autonomous University of Zacatecas and is a specialist in the areas of migration, remittances and regional/rural development. Previously, he was Director of the Faculty of Economics and Academic Secretary of the Autonomous University of Zacatecas. Dr. García Zamora is a member of the National System of Researchers (SNI) Level II; a member of the Mexican Academy of Political Science; advisor to the National Science & Technology Council (CONACyT); advisor to the Faculty of Social Sciences of UAM-Iztapalpa and to the Colegio de la Frontera Norte; an advisor and mentor of the Inter-institutional Doctoral Program in Administration, UAA, UAD, UASLP, UAQ and UAZ, CIDE, COLMEX and the ITAM (Mexico); member of the Sin Fronteras IAP Trust and the INEDIM Trust. Dr. Zamora is a member of the Euro-Latin American Development Studies Network "Celso Furtado"; the International Sociology Association; the Social Economics Network; the Latin America Academic Training (ALFA) program of the European Union (Economy & Society); the Research Consortium on Agricultural Policy; the Radical Political Economists Union; the Frontera Scholarship Association; the Association of Western Social Science. He is a member of the

Sin Fronteras IAP Council; ex-president of the Mexican Association of Rural Studies (AMER); Member of the INEDIM (Institute for Studies and Dissemination on Migration, A.C.) Trust. He has extensive research experience in the field of international migration, migrant organizations and regional/local development.

List of Figures

List of Tables

Ethiopia and Role of Its Diaspora in Development

Introduction

Ajay Dubey and Sushmita Rajwar

In contemporary times, diaspora communities have become influential players in shaping diplomatic and strategic relations between countries, particularly in the realm of foreign policy. These diasporic community not only serve as agents of cultural and social development but also contribute immensely towards economic development in both their host and home nations. This role is exemplified by the policies and initiatives of the Jewish and Chinese diasporas, which have successfully fostered multiple diplomatic connections.

In the twentieth century, the term diaspora came to refer to the dispersal of any ethnic group or community beyond their country of origin. But in today's times, diasporas are widely recognised across the globe, including the Jewish, Chinese, Italian, Greek, Japanese, South Korean, Lebanese, Filipino and Polish diasporas. Since the end of the Cold War, diasporas have played a pivotal role in fostering relationships between nations. They have contributed significantly to both their home and host countries through remittances, philanthropy, skill transfer, investment, advocacy and various forms of assistance.

A. Dubey (✉) · S. Rajwar
School of International Studies, Jawaharlal Nehru University, Delhi, India
e-mail: ajaydubey.sisjnu@gmail.com

A. Dubey and S. Rajwar (eds.), *Diaspora in Development*,
https://doi.org/10.1007/978-981-96-6747-5_1

Through case studies, interviews and data analysis, this book examines how diasporas not only support economic growth but also strengthens social and cultural development, resilience and innovation. The journeys of the various diasporas in the world though different from each other, display a powerful testament to the transformative potential of transnational communities—a testament to how people can shape and support development across borders. As many countries now look to the future, their diaspora will undoubtedly continue to play an essential role in its path towards inclusive, sustainable and globally integrated growth.

With the background of various diasporas and their role in development—social, economic or political in their countries of residence, this book covers a range of issues concerned with the Indian Diaspora. The book includes chapters that talk about the specific policies aimed at the Indian diaspora to connect them with their homelands. The book will discuss the Indian diaspora in general in major chapters while looking at Indian diaspora's role in the development of Mauritius, its significant contribution in the Réunion islands, Indian diaspora's role in the development of Fiji, its social and political contributions in South Africa and also how organisations such as Hindu American Foundation have played a role in promoting Hindu American activism. While the book looks at Indian diaspora in general, it will also discuss other diasporas who have played an important role in development such as the African diaspora in the USA, the Malagasy diaspora, the Pilipino diaspora and the Mexican diaspora in the USA.

Being one of the most significant and dynamic global communities, the Indian diaspora is engaged not only in contributing to the political and economic development of their host countries but also in the development of their home country, India. The Indian diaspora that constitutes almost 35.4 million people (as per the estimate of the Ministry of External Affairs on 21 May 2024) that includes 15.85 million NRIs and 19.57 million PIOs, is now a powerful network of people spanning across the North America, Middle East, Africa and Southeast Asia. This diaspora facilitates economic investment, cultural exchange, knowledge transfer and forms a bridge between India and the World. This book is an attempt to explore the different ways in which the Indian diaspora became a driving force in India's development and also development of their host countries through financial, intellectual and social resources to support progress across sectors.

The role of the diaspora has not been the same in different time periods. During the initial phases, migration was driven by economic hardships, colonial indentureship or academic endeavours. These Indians who migrated to different parts of the world in various capacities, always retained a deep connect with their roots. They kept their identity intact even during the most testing times, and did not allow it to fade away. Post-independence, the Indian economy struggled to maintain itself and did not make any attempts to connect with its diaspora. But the transformation of the Indian economy in the 1990s with the adoption of liberalisation and privatisation methods gave more space for the diaspora to support the Indian economy. It gave many opportunities to the diaspora to connect with India and to invest, partner and engage with India.

Among the various contributions of the Indian diaspora, remittances are one of the most well-known economic contributions. The contribution of the Indian diaspora towards remittances was estimated at a value of US$ 125 billion this year, the highest in the world. But remittances are not the only contribution of the Indian diaspora, it has also contributed to intellectual capital, bringing back knowledge, skills and best practices from various fields. There are various programs like knowledge economy and skill development that have played a crucial role in development.

Indian diaspora plays a critical role in advocacy and policy. By helping to shape a positive international image of India, diaspora communities foster diplomatic and economic ties with their host nations. As the Indian diaspora forms prominent figures in government, business and academia to grassroots advocates and community leaders, these people are a help to project India's interests, strengths and challenges onto the global platform.

While, the contributions of the Indian diaspora are immense, there also lies varied complexities and challenges involved in leveraging these connections for development. Engagement with diaspora requires supportive policies, inclusive networks and recognition of the diverse identities within the diaspora itself. India's policy towards its diaspora had evolved from a policy of disassociation to a policy of active association in the 1990s. Post-globalisation India started framing its policy towards the diaspora and started with initiatives like Pravasi Bharatiya Divas, Overseas Citizen of India (OCI) cards and specialised investment programs, all of which aim to facilitate stronger and more structured connections between India and its global diaspora.

Having realised the immense potential of its diaspora which can be a dominant partner in development, the Government of India adopted various strategies to connect with and leverage its global Indian community. The first chapter titled *Reaching Out to the Diaspora for Development: Some Reflections on the Indian Approach* examines India's evolving policies and outreach efforts aimed at engaging its diaspora to support the country's economic, social, and technological progress.

Historically, the Indian government's interaction with its diaspora was limited, as the government was busy in updating its agricultural sector, removal of poverty, generating employment which were all domestic needs that needed more attention at the time. However, as India started gaining in various sectors like agriculture (green revolution), defence (nuclear tests) and the introduction of economic liberalisation policies in the 1990s, India realised that it was the right time to connect with its diaspora. With a large diaspora settled in various parts of the world and also a fresh wave of migration of professionals during the 1990s, India launched initiatives to facilitate diaspora engagement in key areas such as investment, knowledge transfer and cultural diplomacy. The recommendations of the High Level Committee on Indian Diaspora (HLCID) helped the government to shape its diaspora policy from the early 2000s. Initiatives like the Pravasi Bharatiya Divas (a biennial event celebrating the diaspora), the Overseas Citizen of India (OCI) scheme and the establishment of the Ministry of Overseas Indian Affairs (MOIA) reflect India's efforts to strengthen these ties. The chapter aims to examine as well as critically evaluate the diaspora initiatives of the government.

Indian diaspora in the Indo-Pacific region also constitutes a substantial number and Fiji is one of the countries where many Indians were taken as indentured labour in the nineteenth century. Between the years 1879 and 1916, the indentured labour was brought under the *Girmitiya* agreement and there were other free migrants including Punjabis, Gujaratis, who came and settled in Fiji and hence contributed immensely to the socio-economic development of the country. Fiji is an example where it can be seen how diasporas have been considered agents of change and new actors in the development of the country of origin and the country of residence.

There have been various studies that have examined ways in which countries of origin can benefit from skilled migration through both the diaspora and returnees. It is quite fascinating to know about the three distinct Indian diaspora groups that are present in Fiji including the

Girmitiyas and their descendants, the Gujarati traders and the contemporary free expatriate Indian migrants. The Indians in Fiji have preserved a variety of traditional Indian customs that include the languages, religion and different festivals among them. Different languages like hindi, tamil, Bhojpuri have been maintained through generations. Religions like Hinduism and Islam have also played prominent role among the diaspora as temples and mosques became focal points for community interactions and prayers.

The Indian community has contributed not only in development of art, culture and music but also politics. The political landscape had witnessed a lot of turmoil with the military coup and political instability in 1987 and 2000 that led to emigration of Indo-Fijians to different countries like Australia, New Zealand, Canada and USA. There were many problems that emerged with regard to the land rights, for the Indo-Fijians as their number grew. These issues and complexities have been dealt with in the chapter on *The Diaspora Engagement: Roleof Fiji-Indian Diaspora in Development of Fiji.*

Mauritius is also one of the important countries when we talk about diaspora. It receives huge financial aid from France and also considerable development assistance from India. At the same time, Mauritius recognised how diasporas have been able to bring about the development of this home and host countries. They understood the need of engaging with its own diaspora to drive national development. There being a sizable Mauritian diaspora in France, UK, Australia and Canada, Mauritius started seeing them as predominant agents for knowledge transfer and economic investment.

In order to connect with its diaspora, Mauritius has initiated several programs like the 'Diaspora Scheme', that provides tax incentives to the highly skilled Mauritians to attract investments back to the island. The investments from the Mauritian diaspora have already made significant impacts in the field of real estate, tourism and technology sectors as has been promoted by the government through various schemes. Mauritius has a history of Indian immigrants settled in the country who came in as indentured labour and since have contributed in the society, economy and politics of the country.

Close to Mauritius, the Réunion islands also have a history of Indian migrants who were brought as indentured labour and have contributed economically in the islands. A vast majority of them coming from South India were involved in the cultivation of sugar cane and vanilla both of which were influential to the economy of the island. The Indian diaspora in Réunion islands also established direct trade links between Réunion and India, increasing import and export in spices, textiles and food products. Apart from the economic contribution, the Indian diaspora has also contributed culturally to the region. The Indian festivals and traditions are well-known and the Indians have been able to maintain their Indian identity despite having adapted completely to the French culture. The adoption of many Tamil words in Creole language itself is an example of assimilation of Indians.

It is quite vital for the study of Indian diaspora to examine the contribution of Indian diaspora in countries like Réunion which was a French colony and the population has been assimilated into the French culture. To look at how Indian diaspora has been able to maintain their Indianness is something that draws one towards the uniqueness of the Indian culture. But despite successes, many challenges remain to be addressed. India's diaspora engagement must also balance varied needs across diaspora generations, adapt to regional differences and address social problems around identity and belonging. These issues are discussed in the chapter on *Key Role of Indian Diaspora in Réunion Island Development.*

The general perception of Hindu American activism often highlights a perceived alignment between the actions of Hindu diaspora groups in the United States and the goals of Hindu nationalist organisations in India. For instance, Hindu American Diaspora's influence on the current portrayal of ancient and medieval Indian history in U.S. school textbooks, which they think is misleading and only half-truth, is such efforts that are frequently seen as mirroring similar revisionist campaigns within India. However, while it can be said that there is some connection between Hindu American activism and the Hindutva movement in India, this relationship is more nuanced. Hindu activism in India is a majoritarian movement where the majority of the population belongs to Hinduism, while Hindu American activism operates in a minority context. The next chapter on *Complicating the Narrative of Hindu American Activism: The case of Hindu American Foundation* will examine the Hindu American Foundation's activism, some aspects of which may align with the objectives of the Hindutva movement. The chapter is an in-depth analysis on the Hindu American Foundation, explaining the history and context of the foundation, explaining who is a Hindu American and taking case studies from the members itself, about its ideology and objectives.

Apart from its presence in other parts of the world, the Indian community in the Gulf Cooperation Council (GCC) countries namely Saudi Arabia, the United Arab Emirates, Qatar, Oman, Kuwait and Bahrain constitutes one of the largest groups of migrant workers. They are engaged in various industries, including construction, information technology, finance and healthcare. Their presence significantly influences the economic, social and cultural landscape of their host nations while maintaining strong ties with India. Remittances from the Gulf region have always been a significant portion of the total amount of remittances being received by the Indian diaspora from around the world. However, many of these migrants encounter challenges such as visa limitations, labour rights issues and precarious living conditions. As a result, discussions on policies related to diaspora welfare are essential.

In their host countries, navigating identity and cultural integration presents complex challenges that require deeper academic exploration. Existing research primarily focuses on remittances and labour exploitation, often overlooking the agency, resilience and broader implications of this diaspora for India's global influence. The next chapter titled *Diaspora and Development: The Indian Community's Contribution to the Gulf's Economy and Society* examines the diverse socio-economic and cultural roles of the Indian diaspora in the Gulf, employing transnationalism theory and migration systems theory as analytical frameworks. Migration systems theory highlights the economic, historical and structural factors shaping migration flows, while transnationalism theory explains how migrants maintain simultaneous ties with both their homeland and host country, impacting the economy and culture of both regions. Regulations in Gulf countries generally prohibit migrant workers from organising unionised demonstrations or participating in political discussions. However, informal advocacy groups and social media have emerged as alternative avenues for voicing grievances against these restrictions. Additionally, the Indian government, through its embassies and diplomatic channels, has been involved in efforts to support and advocate for the rights of its migrant workforce. Enhancing the well-being of the Indian migrant community in Gulf nations, especially those in low-wage jobs, requires urgent policy changes. One critical area that demands immediate attention is the oversight and regulation of recruitment agencies to ensure fair and ethical labour practices. This chapter sheds light on evolving migration trends and underscores the crucial role of the Indian diaspora in driving economic, social and cultural transformations across the Gulf region.

Indians also have a long history of migration to South Africa mostly as indentured labour. Besides, India has been a country that has rallied against the racial discrimination in the form of Apartheid in South Africa, at every multilateral organisation including the UN and NAM. Indian foreign policy was quite limited during the time of the first Prime Minister Jawaharlal Nehru who suggested that Indians must work with their African brethren, they must fight alongside them and support their struggles for independence.

During the late 1990s, the creation of High-Level Committee on Indian Diaspora brought about a complete U-turn in the earlier policy towards Indian diaspora. However, there were some key measures taken up to promote NRI investment in the country during the 1980s also. But from early 2000s, the introduction of measures like Pravasi Bhartiya Diwas, Pravasi Samman, OCI card India have made numerous efforts to engage with the old (PIO) and the new (NRI) diaspora as well. The chapter on *Diaspora and Development: A Study of Indian Diaspora in South Africa* traces the history of the Indian migration to South Africa, participation of Indians in the fight against racial discrimination, awakening of the political consciousness through the efforts of the Indians in South Africa and the combined struggle for independence by Indian South Africans as well as the Africans. The chapter is also a detailed reflection of India's foreign policy towards its diaspora that has evolved from a period of complete disassociation to a phase of active association with the help of various new measures.

The engagement with the Indian diaspora by the GOI follows a model which reflects a mutual understanding aimed at the development, at establishing long-term connections and also building networks among the Indian community that exists globally. This paper is an excellent analysis of India's Diaspora policy and its role in strengthening relations between India and South Africa as it seeks to examine the contributions of the Indian diaspora not only in their home nations but also in their host nations. Interestingly, the paper highlights various initiatives taken up by the Indian Diaspora during the COVID-19 global pandemic.

Besides the diasporas in the African countries, there are many Africans also who have migrated from their countries to other parts of the world. These groups from a good number of those that came to be recognised as the African diaspora but had their affiliations towards their respective African countries from which they migrated. Africans who have migrated from Madagascar form a substantial number of the international migration history. Notably, the Malagasy diaspora is largely composed of skilled or semi-skilled individuals, making this human resource a potential driver for the country's socio-economic development. This diaspora is spread across different parts of the world, with notable communities in France, the United States, Canada and parts of Africa. France, being Madagascar's former colonial power, hosts one of the largest Malagasy communities due to historical, linguistic and cultural ties. Despite their understated visibility, the impact of their activities frequently garners attention in the media, often showcasing tangible and significant outcomes. Whether acting independently or through collective associations, the Malagasy diaspora demonstrates remarkable dynamism in contributing social and financial remittances to support their compatriots in Madagascar. However, these contributions primarily occur outside the framework of government initiatives. The chapter on *Malagasy Diaspora: Translucent but Palpable in its Engagement* focuses on the diaspora engagement policies of the government which started with the establishment of the directorate in charge of the Malagasy diaspora and examines how this policy has reinforced the link between diaspora and Madagascar.

While discussing the character and strengths of different diasporas settled in the United States of America, one more dominant diaspora other than the African is the Filipino diaspora. The first wave of migration to the USA from Philippines was mainly skilled labour migration in factories and military facilities. During the 1960s, there was more migration of doctors, nurses who support the booming economy of the USA. At the same time period there was an influx of bright students under scholarships and exchange programs at various universities in USA, who excelled in academics, later gained employed and settled in the USA. Another wave of migration occurred in the 1970s when the government of Philippines itself encouraged migration in order to balance against the high rate of unemployment in the country. The migration occurred not only to USA but also to the Middle East for skilled professionals in the petrochemicals as well as infrastructure.

With the increase in migration from Philippines, the government started making its efforts to engage with the diaspora by setting up Boards and Councils to take care of the overseas employment of Filipinos. The policy of the government has focused extensively on providing specialised assistance to its diaspora which have been chosen carefully after assessment of the needs of its diaspora. The establishment of the Global Filipino Diaspora Council and the Diaspora 2 Dialogue (D2D) program that includes atleast 10 areas of interventions ranging from education, arts and culture, business, tourism to medicines and even return of the diaspora. The various global summits for diaspora have proved to be quite fruitful and productive in the engagement of the diaspora. By discussing the case study of Filipino diaspora, the chapter on *The Private Sector, Diaspora Engagement and the Migration-Development Nexus: Comparative Cases from the Philippines* draws one's attention to the various diaspora engagement policies of the governments that is trying to maintain the mutual benefits of all three—the home country, the host country and the diaspora.

The migration from Africa in the form of transatlantic slave trade has a dark history, which began after the arrival of the Europeans in the continent and forcibly brought millions of Africans to Brazil, Europe and USA. Over the centuries, this connection significantly shaped Africa's relationship with different regions. After the abolition of the slave trade, the influence persisted as many freed African descendants chose to return to Africa, while others remained in their migrated countries, integrating into society and contributing to its economic, political and cultural fabric. Therefore, the study of African diaspora in other parts of the world requires a special attention and research. These migrated Africans, many of them who were forced to migrate due to the slave trade and many who went after the independence of their home countries, have a distinct relationship with their countries of origin. Their home countries have tried to connect with their homelands and also contribute in the development of their home nations through various methods.

The most discussed migration among the topic of diasporas in the USA is the migration of Mexicans to the USA. Mexican migration to the United States has been a significant and complex phenomenon shaped by historical, economic, social and political factors. After the U.S.-Mexico War (1846–1848), the Treaty of Guadalupe Hidalgo (1848) led to the transfer of large Mexican territory to the USA which meant that many Mexicans became USA citizens overnight. In the following decades, there began a migration of workers towards the north for work in mining, agriculture and railroads. During the Mexican Revolution (1910–1920), political and social upheaval prompted many Mexicans to seek stability and employment in the United States. The demand for labour during World War I (1914–1918) further increased Mexican migration. Mexican migrants have settled in the USA and established their own groups and clubs which are mostly based on their hometowns. These clubs have been engaged in various activities like philanthropy that includes some activities like maintenance of local churches, gardens and parks. The community development projects in the 1990s by these clubs have made remarkable contributions for infrastructure where the federal and state governments also contributed an equal amount that the migrants contributed. The programs expanded further during the twenty-first century when the local governments also started contributing, resulting in the huge success of these 2 × 1 and 3 × 1 programs.

However, there are many economic challenges that have caused unemployment in the recent years resulting in migration. These migrant organisations also need a lot of strength in terms of their structures and this would require increased investments from migrants and alliances with various sectors. The chapter on *The Mexican Diaspora in the US and the 3 × 1 Program as Transnational Community Development: Lessons and Challenges* talks about the various contributions of the Mexican diaspora in the USA and what are the challenges they face in sustaining their community development programs.

Ethiopian diaspora is also a very interesting case study to understand their role in the development of their homeland. It is estimated that more than 2.5 million Ethiopians currently reside outside their homeland, a result of continuous migration patterns throughout history. Historically, Ethiopians have relocated for purposes such as trade, religious missions, scholarly pursuits and diplomatic engagements. After World War II, a significant number of Ethiopian students and intellectuals sought educational opportunities abroad, particularly in Europe and North America. According to a 2014 report by the Migration Policy Institute, 60% of Ethiopian immigrants in the United States arrived after the year 2000. The Ethiopian migrant population in the United States grew significantly, rising from 35,000 in 1990 to over 220,000 by 2017.

With globalisation and economic shifts, the Ethiopian diaspora has become an important source of foreign currency and remittances. Their involvement has had profound effects on Ethiopia's domestic and international political and economic landscapes. The widespread Ethiopian community worldwide has also influenced Ethiopia's diplomatic relations with various nations. The diaspora plays a crucial role in shaping the country's foreign policy, contributing to economic, political and cultural exchanges. Many Ethiopians abroad view themselves as active participants in Ethiopia's democratic evolution. In regions such as North America and Europe, the diaspora has played a pivotal role in shaping Ethiopian political discourse, both positively and negatively. Additionally, they have actively engaged in promoting the values and societal structures of their host countries within Ethiopian communities. With this background the last chapter titled *Ethiopia and Role of its Diaspora in Development* focuses on the role of Ethiopian in the development of Ethiopia, their economic contributions as well as their political influence. The chapter discusses in detail about the engagement policies of the Ethiopian government as well as significant changes it has brought towards establishing that connect with the diaspora.

Therefore, one can see that diasporas especially Indian diaspora has been a transformative force in the development journey of its nation, where it has been able to showcase its unique ability to bridge cultural, economic and political divide. With almost 50% of the Indian diaspora in the world accounting to the 'old diaspora' or the 'Girmitya' or the 'indentured labour' diaspora that went to countries of South Africa, Mauritius, Fiji, Caribbean and others, Indian government tailored its specific policies to cater this old diaspora. Indian government could very well understand that this is soft power that exists in large parts of the world and must be engaged by the home government. While on the other hand, there were policies also to engage with the 'new diaspora' that could prove to be more beneficial in transfer of knowledge and skills that drive innovation and growth. This global diaspora has also played a key role in the enhancement of India's global image, which helped India to put forward its own interests in international forums and even strengthening its bilateral relations with the host nations. The proactive engagement of India through its various policies like PBD have also encouraged investments and collaborations that have further amplified the impact of its diaspora. As India continues to reach its aspiration to become a global power, its diaspora is going to remain its invaluable partner in shaping India's development narrative.

Reaching Out to Diaspora for Development: Some Reflections on the Indian Approach

Chandrashekhar Bhat

The Context

The concept of globalization acquired much broader meaning and application than mere movement of capital and markets across several countries towards the end of twentieth century. According to Lechner (2005: 330) the process of globalization today covers '...worldwide diffusion of practices, expansion of relations across continents, organization of social life on a global scale and growth of shared global consciousness'. Globalization has lead to increasing interdependence between countries, not only in terms transfer of capital and goods but in almost all spheres of life, from lifestyle, consumer behaviour, fashion to ideology and spiritualism. With the revolutionary advancement especially in technologies of information, communication and travel during the last two decades of the twentieth century, there has been a steep increase in the global movement of capital, technology, cultural commodities, media, cuisine, people, deities and modes of worship across several nations.

Globalization is so complex a process today that no theory can adequately offer its multifaceted manifestations. Hence scholars have

C. Bhat (✉)
University of Hyderabad, Hyderabad, India
e-mail: csbhat.csid@gmail.com

A. Dubey and S. Rajwar (eds.), *Diaspora in Development*,
https://doi.org/10.1007/978-981-96-6747-5_2

emphasized on different perspectives to conceptualize globalization. For instance, Robertson (1992) looked at the process from a cultural perspective, illustrating how the world becomes shrunk and awareness of the world as a whole increases. The emerging organizational and technological developments, according to Robertson (*ibid.)*, have lead to the miniaturization of the world and construction of a global field in which individuals and societies become part of a larger system of societies globally dispersed. According to Robertson (ibid), globalization produced a more integrated yet also differentiated world society. He popularized the concepts of 'glocalization' and 'localization' to emphasize the incorporation of some of the global or universal features in local cultures and also the incorporation of local features at the global level. Taking the local specificities from culture, commodities, cuisine, music and so on to the global reach is of particular interest to diaspora communities.

Contemporary globalization is augmented by 'sociotechnological infrastructure'—social uses and application of advanced technologies of communication and transport—increasing the magnitude of international migration, transfer of commodities, ideas and capital more than ever (Nonini 2000: 565). Appadurai (1990) captures the dynamics of diasporic dispersal through terms such as 'global flows' or *scapes.* He uses *ethnoscapes, technoscapes, financescapes, ideoscapes* and *mediascapes* to analyze the contemporary nature of the movement of people, transfer of technology, capital, ideology and media that celebrated the local in the global scape.

Robin Cohen coined the word *diasporization* to emphasize the role of diasporas in the process of contemporary globalization. 'Globalization and diasporization are separate phenomena with no necessary causal connections, but they "go together" extraordinarily well' (Cohen 2008: 154). For him (*ibid:* 155), globalization has enlarged the practical, economic and effective roles of diasporas, showing them to be highly adaptive arrangements of social formations. There is a greater awareness presence of global civil society, dual/multiple or global citizenship, global governance, governmental and non-governmental multinational organizations which have emerged as institutional arrangements of this emerging transnational or global society.

'Power of Diasporas', as Cheran (2004: 7) mentions, is the significant influence that diasporas or transnational communities are assuming today in political processes, economies and wealth creation at the local, national and global levels. Diasporas are emerging as leading players in global transactions, be economic or political or cultural.

India today is not confined just to a territorial nation-state but extends far beyond wherever Indians are dispersed and made their homes, covering almost all corners of the globe. Indians have taken and embedded 'India' with them to the far off Caribbean Islands of Trinidad and Tobago, Guyana, Surinam, Guadeloupe and Martinique, to the Pacific Island of Fiji, to all corners of South East Asia and Africa, and to North America and Europe. As her people get widely dispersed under the forces of twenty-first century globalization, the new policy initiatives of envisage bringing all Indians together for inclusive development partnering with her diaspora. In other words, under the reign of contemporary globalization, India stretches to all corners of the world where Indians remain and cherish Indianness, forming '*Global Indian Family*'. Globalization today has given rise to a huge transnational social space where Indian families and organizations extend their family, caste, linguistic, regional or national ties across the globe, that the Gottingen scholars (Hermann and Loimeier 2012) most aptly entitled the international conference they hosted 'India beyond India' (Bhat 2012).

Diaspora and Development—The Precursors

Among the several global diasporas, the Jewish, Armenian and Chinese diasporas are briefly examined here as forerunners for their early initiatives to involve in the overall development of their motherland, setting the trend of vibrant diasporic intervention for advancement since the mid-twentieth century. Foremost among the well-established diasporas is the Jewish diaspora, which has a long history of dispersal spanning over two millenniums and more, followed by the Armenian diaspora. In both the cases, the diasporas far exceed the present population of Israel and Armenia, respectively. China has one-third of her population in the diaspora and she has already established linkages with her 50 million strong diaspora to promote mutual interest in socio-economic and political development. By the 1990s the wealth of the Chinese diaspora in terms of their economic output of about $450 billion was almost equaled to that of mainland China (Seagrave 1995). A brief note on diaspora and development follows, illustrating the participation of Jewish, Armenian and Chinese diasporas in the advancement of their motherlands.

The Jewish Diaspora

Diaspora as a resource has been first realized by Israel right from her inception till this day. The emergence of the new nation-state of Israel (May 15, 1958), following the UN resolution for the establishment of two states in Palestine despite the British and Arab resistance, and her development since then to be a significant global player has been extensively promoted by the Jewish Diaspora world over.

Global Jewish population is estimated to be nearing 13.8 million with over 6 million in Israel and nearly 7.8 million in the diaspora. It is important to note that the population of Jews in the US is closer to the current population of Jews in Israel (DellaPergola 2013: 5). According to Safran (2012), the diaspora's involvement in overall advancement of Israel has continued in many ways, such as contributions to the Jewish National Fund, the Hebrew University, a variety of other Israeli institutions as well as research activities, exchange programs, lobbying with their governments and legislatures besides philanthropic Zionism. Israel also acknowledges the continuation of the diaspora as a centre of Jewish society and state.

The Jewish diaspora of the US has been instrumental in providing Israel with necessary support—diplomatic, economic and military—almost amounting to more than $115 billions from 1949 onwards. There is extensive collaboration and dependence in defense production and security matters, including terror tracking, between the two countries. In fact, their mutual partnership benefited both Israel as the US. It is widely believed that in the field of information technology partnering of US companies with Israel was instrumental to Silicon Valley's success. In fact many of Intel's most successful microprocessors, accounting for nearly 40% of the firm's revenues, were designed and developed in the company's research and development centres in Israel. Israeli IT researchers have significantly contributed to the invention of a secure mode of financial transaction through an instant message on the Internet, or online shopping using PayPal (Eisenstadt and Pollock 2012).

The Armenian Diaspora

Armenia was under the occupation and rule of Soviet Union for nearly seven decades from 1915 until the declaration of independence on August 23, 1990 and became Republic of Armenia on 16 October 1991. Armenia was the first country to adopt Christianity as the state religion as early as 301 AD but was later subordinated by Islamic forces, establishing an 'Emirate of Armenia' and by Turks during which a large number of Armenians migrated to the neighboring territories. Like the Jews, Armenians too have gone through similar experiences of genocide and exile for centuries. The last invasion was by Russians, who split the country into Turkish, Persian and Russian Armenia in 1828. The attempt by all Armenians, through nationalist and revolutionary mobilizations, to reunite Armenia was fiercely opposed by Ottoman Sultan killing more than 300,000 and forcing large-scale dispersal from Turkish Armenia during 1894–1896. During World War-I the Turkish forces, allied with Germans, carried out the worst traumatic massacre of nearly a million Armenians. An equal number out of those who survived the atrocities fled to the Middle-Eastern countries like Lebanon, Syria, Palestine and Iran besides Europe and US. Today the Armenian diaspora in Russia (2.3 millions), the US (1.5 millions) and France (400,000) is considered to be both populous and highly resourceful (UK Foreign and Commonwealth Office 2015). The full potential of Armenian diaspora is yet to be realized by the homeland initiating inclusive policies for mutual growth (Manaseryan 2004).

In the US, for instance, Armenians have a strong political lobby in Washington and often join the Jewish-Americans in sponsoring joint exhibitions, conferences and publications. There are several well-established Armenian diaspora organizations such as the Armenian Assembly of America, Armenian General Benevolent Union, Armenian Historical Research Association, the Zoryan Institute, the Hairenik and Baikar Associations of Boston, the Armenian Relief Society, and many more to provide economic, educational, political support and relief to Armenians and Armenia. Wealthy Armenians in the US undertake philanthropic activities covering health care, education and other developmental programs in Armenia. Since 1991 Armenia has received several billion dollars' worth of financial support from the US government besides prominent donors like the Lincy Foundation, founded by California-based millionaire Kerk Kirkorian, has invested nearly $300 million in Armenia since her independence. There are as many as five Armenian newspapers in the USA and most of the Armenian churches are promoting schools, cultural and recreational centres (Cohen 2008).

Armenian diaspora is much larger than the population of their homeland. The recent Armenian population estimates show 3,060,631 while the Armenian diaspora is estimated to be around 9 to 10 millions. Armenian diaspora is almost three times the Armenian population today and form a very significant resource for Armenia's development (http://www.indexmundi.com/armenia/demographics_profile.html downloaded on 02.01.2016).

The Chinese Diaspora

The Chinese Diaspora is perhaps the next one to follow in transferring her resources for the advancement of her mother/ancestral land. The most populous country of the world with nearly 1.4 billion population has the largest diaspora of 50 millions dispersed densely in Southeast Asian countries. Traditionally the overseas Chinese were believed to pay back to the community and locality of their origin, which shall be the place for their eternal resting. There are accounts describing how the overseas Chinese immigrants made special efforts to take/send the bodies, even bones, of the dead to their native villages for burial via London where there was provision for keeping corpses during the transit.

The phenomenal rise of Chinese diaspora since 1950 followed the construction of what Kuznetsov and Sabel (2006: 13) called '*bamboo network*' connecting all Chinese from Hong Kong (now under China), Malaysia, Singapore, the Philippines, Indonesia, Taiwan and Thailand to one another and also to China. These networks included family firms operating in traditional trading and manufacturing and then later extending them to the emerging domains of finance and high-tech firms. Once they start an enterprise in one host country, they expanded and diversified to include neighboring countries. The overseas Chinese extended their businesses, supporting their family members to invest in new ventures in other promising locations. The new firms grew on the founding family's capital, social and political networks. Several family firms grew into huge business houses, controlling a variety of small and medium firms in different countries. Overseas Chinese emerged prominent in trade and businesses pertaining to real estate development, construction and manufacturing industries, depending particularly on their ability to combine traditional skills with modern technology.

Chinese diaspora came much closer to mainland China with the open door policy of Deng Xiaping in 1978 who saw a huge human and capital resource among the overseas Chinese that can be gainfully tapped for Chain's advancement. Hong Kong and Taiwan were strategically located in terms of cultural and geographical proximity besides historical ties. Several overseas Chinese multinational companies made inroads into mainland China and diversified manufacturing with low-cost production for the huge international market. China discovered the need to partner with her diaspora and provide necessary guidelines for investment and philanthropy (Kuznetsov and Sabel 2006). It is common today to find the wealthy overseas Chinese investing in the place of their origin for philanthropic endeavors as they believe in the family and ethnic loyalties.

India and Indian Diaspora

That the Indian diaspora is perhaps the most diversified among the diasporic communities in the world cannot be overstated and one of the recent Oxford University publications (Jayaram 2011) is exclusively devoted to *Diversities in the Indian Diaspora* and India too is differentiated in terms of racial, linguistic, regional, religious and caste diversities.

The foremost among the diasporic diversities is the distinction between the *PIO* and the *NRI* diaspora. 'Not only do they vary in the contexts of their emigration and destinations but also in terms of their socio-economic background and the degree of interaction with the motherland. While the *NRI* Diaspora has retained vibrant relationship with their families, communities and localities of origin in India, majority of the *PIO* Diaspora have lost their contact with the motherland. On the other hand, in course of their long journeys by ship to distant destinations, the unknown co-passengers became "*jahaji bhai*" (literally meaning "*ship brother*", a brotherly affinity owing to sailing together) losing their caste and religious identities to a large extent, including their mother tongue' (Bhat 2006). They had no access even to their own kith and kin working in other plantations under the '*new system of slavery*' (Tinker 1974) invented by the British colonialists called 'indenture labour'. Of course, they had no way to communicate leave alone to travel back to engage with the kith and kin left behind in their places of origin until the period of indenture contract was over.

The Indians who emigrated after India's independence—the *NRI* diaspora—had certain advantages of professional education and training besides belonging to Anglophone urban middle classes hailing from upper castes. Employment in industrially developed countries offered them adequate income that could facilitate frequent visits and communication with the places of their origin. There have always been instances of mutual economic, educational and cultural support in times of need between *NRI* diaspora and their families and places of origin in India. Though to a limited extent, some of the successful overseas Indians have been 'giving back' through philanthropic activities at their places of origin, in terms of health care, education, rural development and so on. However such efforts in promoting economic and social development have escaped any state or national attention till recently (Bhat 2003).

The NRIs and NRI Diaspora

The Government of India ignored the presence and concerns of overseas Indians until almost 1990s. When India was going through foreign exchange reserves crises, the government realized the potential of the successful NRIs and NRI Diaspora, who had emigrated during the later half of the twentieth century to the developed world, especially to the USA, UK, Canada and Australia), Saudi Arabia, UAE and few other Gulf countries. They came from the middle-class urban families and had acquired necessary professional skills to be absorbed as doctors, engineers, scientists and software professionals. The new industrial policy introducing liberalization and privatization in 1990s sought the involvement of overseas Indians in investing both in terms of capital and technology. The decade saw the nomenclature of NRIs assuming significance in enhancing India's economic growth. With the view to promote the interest of overseas Indians, a Global Organization for the People of Indian Origin (GOPIO) was established at the International Convention in New York in 1989, which also initiated dialogue for support with the Government of India. Around this time India was going through severe crises in Balance of Payment and in order to overcome this, the Government of India—through State Bank of India—floated 'India Development Bonds (1991)', 'India Resurgent Bonds' (yielding $4.2 billion in 1998) and India Millennium Deposit (2000), issued exclusively to overseas Indians (both NRIs and NRI Diaspora), which were over subscribed. NRIs were encouraged to invest in India through certain attractive schemes as much as they were invited to set up industries and encouraged to transfer technology to promote India's progress.

The NRIs world over, especially in Gulf countries, have immediate or part of their families and interest back home at the place of origin. A major portion of their hard earnings is remitted back home for family maintenance, healthcare, education and asset creation (like property, housing, business etc.) as they can never hope to settle down or make long-term investment in those countries. India is perhaps the recipient of the largest amount of remittances that provide considerable resources for development. In India it is the state of Kerala that received the major share of remittances from Gulf countries. 'Remittances from emigrants abroad to Kerala in 2011 were estimated to be approximately Rs. 49,695 crores compared with Rs. 43,288 crores in 2008..... Increase in remittances during 2008–2011 (15%) was much larger than increase in the number of emigrants (4%)' (Zachariah and Irudaya Rajan 2011: 6). Kerala received remittances to the extent of 31% of the state's Domestic Product.

According to the South Asian migration study, conducted by the Centre for Development Studies in Thiruvananthapuram, the '...annual remittances to Kerala in 2004 itself had reached to the extent 1.74 times the revenue receipts of the state, 1.8 times of the annual expenditure of the Kerala Government, 7 times of what the state received from the Central Government, and contributed nearly 22% of the Net State Domestic Product' (Zachariah and Irudaya Rajan 2004). Remittances from the gulf migrants has led to a notable level of development in sectors like housing, education, healthcare, transportation, communication, rural development and several other infrastructural facilities.

The NRI diaspora, on the other hand, keeps major part of their earnings in order to spend on their transport, skill enhancement, housing, children's education and invest if they can save further. In a few instances, they may remit for the maintenance of parents and/or siblings in India. Those successful professional and entrepreneurs may invest in their Indian undertakings or remit for philanthropic activities, in places or regions of their origin. States like Gujarat, Punjab and Andhra Pradesh have been receiving such investments and developmental projects besides the contribution for development from NRI diaspora philanthropists and NGOs. I shall be discussing these in detail later.

The PIO Diaspora

The People of Indian Origin Card was launched by the Ministry of Home Affairs in March 1999 as a response to the emotional bonds of *PIO* diaspora who have made other countries their homes since a century or more, but now desire to renew their ties with the ancestral land of their origin.

Those Persons of Indian Origin (PIOs), up to the fourth generation, settled anywhere in the world except for a few specified countries (e.g., Bangladesh, Pakistan) are eligible to avail themselves this facility of PIO Card. The foreign spouse of a citizen of India or NRI Diaspora would also be covered under the scheme. This scheme will entail all facilities to PIOs which were generally open to Non-Resident Indians (NRIs).

Except agricultural or plantation properties, the PIO Card holders are entitled to buy, hold, transfer and dispose of immovable properties in India and their children may be admitted in all educational institutions in India under the general category quota for NRIs and the PIOs may participate in all housing schemes including those of government agencies like LIC. The PIO Card holders need not get a visa to visit India and also not required to register with the Foreigner's Registration Officer if the continuous stay of less than 180 days. The PIO Card was especially designed to enable the visit of people of Indian origin to their roots in India and to facilitate their stay and visits to pilgrimage centres hassle-free.

The last decade of the twentieth century is marked by success stories of noteworthy achievements by Indians in several advanced countries besides the Silicone Valley (of California) in particular. They had excelled in science and technological enterprises, engineering and medical professions and in the newly evolving software industry. India saw them as a resource at a time India was going through difficult times financially. The Government of India realized the potential of Indian diaspora that could be partnered in India's development. In the meanwhile several global organizations of the people of Indian origin, NRI Diaspora and NRIs lobbied for recognition and new policy initiatives to reconnect Indian diaspora with India.

The Ministry of External Affairs, Government of India, appointed a High Level Committee on Indian Diaspora was in September 2000, with the approval of the Prime Minister Atal Bihari Vajpayee, to recommend a broad and flexible policy framework for all overseas Indians, after reviewing the status, needs and role of People of Indian Origin (PIO Diaspora) and Non-Resident Indians (both NRIs NRI Diaspora).

High Level Committee on Indian Diaspora (HLCID)

The HLCID Committee was appointed with Dr.L.M.Singhvi, Member of Parliament (MP) and former High Commissioner of India to UK as its Chairman. Shri J.C. Sharma, Secretary (NRI & PV) in the Ministry of External Affairs, was appointed as the Member Secretary. Members included Shri R. L. Bhatia, MP and former Minister of State for External Affairs, Shri J. R. Hiremath, Indian Foreign Service (Retd.), and Shri Baleshwar Agarwal, Secretary General of the Antar Rashtriya Sahyog Parishad. One of the important terms of reference to the Committee was to study and recommend the contribution that the PIOs and NRIs could make in the economic, social and technological development of India. The Committee was also mandated to look into the facilities, privileges and rights extended to them. The Committee was asked to examine the conditions of the existence overseas Indians including their rights and discrimination, if any, in the countries of their residence.

The Committee took two years of travel and deliberations in India and abroad and submitted its report in January 2002 to the Prime Minister, recommending several measures to resolve the problems faced by NRIs and the PIOs and suggesting to take up '...country-specific plans for forging a mutually beneficial relationship... and for facilitating their interaction and participation in India's economic development' (MEA 2002: iv). The Global Organization of People of Indian Origin (GOPIO), which has been holding Global Conventions both in India and the countries where Indian diaspora has considerable presence, actively participated with the HLCID during their deliberations to evolve significant diaspora policies for mutual advancement in the New Millennium.

The members of the Committee (HLCID) have extensively interacted with the Indian Diaspora community and their leaders and representatives in different countries where people of Indian origin have their significant presence. They have come out with several recommendations, addressing all the four streams of Indian Diaspora, for the consideration of the Government and implementation. Any policy aimed towards Indian Diaspora communities begins with the recognition that the overseas Indians are essentially a product of Indian socio-political reality, from the historical past to the present. The people of India and the Indian diaspora differ essentially in terms of nationality and citizenship. Hence, while formulating the new initiatives the Committee (HLCID) believed that all persons of Indian origin are made to feel at home in India in all respects.

Recommendations of HLCID

The High Level Committee on Indian Diaspora (HLCID) recommended the government to respond to some of the significant issues that had figured in their discussions with the members of the diaspora communities in its interim report. These included (i) a lower fee for PIO Card with 10 years of validity (than 20 years of validity for $ 1000), (ii) clebration of *Pravasi Bharatiya Divas* on January 9th (marking the day Mahatma Gandhi's return from South Africa) of every year, both in India and diaspora countries, recognizing the role of Indian Diaspora in the promotion of India's interest, (iii) award of *Pravasi Bharatiya Samman* to eminent PIOs and NRIs and (iv) *Pravasi Bhavan*—a single window for diaspora dialogue and data bank on Indian diaspora.

The HLCID has examined important issues pertaining to Indian diaspora, such as promotion culture, education, media, economic development, healthcare, science & technology, philanthropy and dual citizenship with the view to recommend appropriate steps to bring Indian diaspora closer India in an institutionalized manner. The Committee has recommended setting up of an empowered body, similar to *Planning Commission* and a *Standing Committee of the Parliament* to monitor interaction between India and Indian diaspora.

To promote Indian culture and mass media, the Committee suggested the involvement of both the people and government of India need to undertake effective steps to *promote languages, philosophy, spirituality, performing arts and ethnic Indian media* from the places of origin among the Indian diaspora communities, especially those of the PIO Diaspora. Of course, the people of the NRI Diaspora still continue to maintain strong family linkages and community bonds in the motherland. Various programs and activities through which Indian culture and media can effectively reach the younger generation among the diaspora have been elaborated in detail by the HLCID in their final report.

The Committee believed that the active participation of people of Indian diaspora and India besides that of the government of India are essential for the effective implementation of various recommendations in the fields of education, economic development, healthcare and science & technology, etc. Recommendations in these areas included a detailed list of the mutually beneficial issues to be articulated and also specifying the facilitating role that the Government of India needs to undertake.

Underlining the emerging significance of philanthropy in reaching out to the people and places of origin, the HLCID in its Report (MEA 2002) suggests that the Government of India and also all the *state governments* to remove all the obstacles for *promoting philanthropic and voluntary or welfare activities of NGOs* that the members of Indian diaspora wish to pursue in India.

Dual Citizenship

Another important issue articulated particularly by the NRI diaspora before the HLCID was that of *dual citizenship*. The demand for dual citizenship was forcefully put up by the prominent members of Indian community in North America and a few other advanced countries during their meetings with the Committee. They argued that the provision of dual citizenship would facilitate their *seamless* participation in India's growth through investments, trade, tourism and philanthropic contributions. Under the Citizenship Act of 1955, all Indians have to give up Indian citizenship when they acquire the citizenship of a foreign country. The HLCID decided the issuing of 'dual citizenship' can only be taken up after ascertaining appropriate safeguards to India's security concerns and also after amending the Citizenship Act of 1955. However, the Government of India decided in 2005 to issue 'Overseas Citizen of India (OCI)', in lieu of Indian citizenship, to those who had forfeited Indian citizenship in the recent past. However, the HLCID denied holders of such citizenship any participation in the electoral process or civil services in India.

The First Pravasi Bharatiya Divas (PBD) bringing together Indian diaspora and India—under the logo of *The Global Indian Family*—was hosted during January 9–11, 2003. Since then these annual conventions are being held every year coinciding with January 9th. So far there have been 13 PBDs organized, 6 in Delhi and the rest in different state capitals. The PBD celebrations are often organized during January 7 to 9 and are always inaugurated by the Prime Minister of India, along with a Chief Guest, generally a PIO Prime Minister or President. The Prime Ministers of India used this occasion to announce new policy initiatives. A couple of plenary and parallel sessions were held during which several union ministers or state ministers holding charge of finance, commerce and industry, health, science and technology, tourism, etc. are invited to address the delegates specially drawing their attention to the vast opportunities available for participation and investments. An entire morning parallel session is organized to showcase the opportunities in each state and several state Chief Ministers make it point to address and invite the NRIs, NRI Diaspora and PIO diaspora to their respective states for global summits, soon after the PBD, offering them incentives for investments. Some of the prominent states that have been participating in this program, year after year are Gujarat, Rajasthan, Punjab and Andhra Pradesh. These states also have pro-active NRI/PIO Departments to facilitate and coordinate with diaspora entrepreneurs.

The President of India always gave the valedictory address after presenting *Pravasi Samman* to half a dozen distinguished achievers among NRIs and Diaspora communities. Cultural evening offered Indian classical dances and Bollywood events. Delegates ranging from 1,500 to 2000 came from 15 to 25 countries and some of the countries, such as the US, Malaysia and Singapore, Mauritius and Reunion, South Africa and Sri Lanka, have been mobilizing large delegations ranging from 25–100 members. About 25 to 30% of the delegates attended PBD every year and continued beyond by a couple of weeks visiting the places of tourist or business interest, relatives and friends and places of origin.

The NRIs and diaspora entrepreneurs converge at their home states soon after the PBD convention to pursue business and investment interests. The states organize their 'Global Summits' generally in association with Indian Chamber of Commerce and Industry or Confederation of Indian Industry state chapters. In fact it is in these summits that the developmental projects are finalized for implementation or business deals are finalized.

It is a great initiative that the Government of India has undertaken to bring Indian diaspora closer to India, officially recognizing the presence of 25 million strong Indian Diaspora and to lay a strong foundation for enduring linkages between India and the Indian Diaspora through formulating several new policies. The High Level Committee has made recommendation in all spheres of life for approval and implementation by the Government of India. Since India's independence, it is a path breaking initiative to bring the Indian Diaspora closer to India.

In order to exclusively deal with the Indian diaspora, the Ministry of Foreign Affairs appointment for the first time an official of the rank of *Additional Secretary* in 2000. The government a step further to create a new ministry in 2004—*Ministry of Overseas Indian Affairs* (MOIA)—to take up all matters relating overseas Indians and approve and implement recommendations of the HLCID and also take up new initiatives from time to time.

New Initiatives Implemented

1. Supernumerary quota of 15% seats in all the institutions of higher education/Universities to the children of Gulf migrants, with fees on par with Indian students.
2. Transport arrangement for Mortal Remains of workers in the Gulf countries
3. Establishment of Indian 'Community Welfare Fund (ICWF)' for workers in Embassies.
4. Know India Programme is a three-week orientation program hosted for diaspora youth by the Ministry.
5. The Study India Programme for diaspora youth.
6. Overseas Indian Youth Club (OIYC) with help of Indian Missions.
7. Scholarship Programme for Diaspora Children (SPDC) for 100 youth annually @ $4000/annum for pursuing undergraduate programs.
8. 'Tracing the Roots' scheme to help PIOs in locating their roots in the places of origin.
9. *Mahatma Gandhi Pravasi Suraksha Yojana (MGPSY)* 2014—a Social Security Scheme for overseas migrant workers with 'Emigration Clearence Rrequired' passports—to provide insurance cover in the event of death or permanent disability; also the scheme ensures old-age pension and helps workers plan for their settlement following their return to India.
10. OCI and PIO Cards *merged and the VISA for life*, 2015
11. *Voting rights to NRIs*: The Election Commission of India to arrange facilities, with the help Indian Embassies and Consulates, for NRIs' voting in Indian elections, 2015 (*It may not be far to grant dual citizenship, as demands are already articulated, with voting rights not just to NRIs but also to NRI Diaspora!*).

Several initiatives to constructively engage the diaspora for sectorial development in different states of India, from where the diaspora had emigrated, were undertaken as early as 1980s. Some of the pro-active states, such as Andhra Pradesh, Gujarat, Kerala and Punjab had already provided certain incentives for diaspora participation in the promotion of industries, trade and business through the local chapters of Indian Chamber of Commerce and Industry (FICCI), Confederation of Indian Industry (CII) and several state departments dealing with tourism, transport, healthcare etc. The Industrial Policy of 1991 offered added advantages to the diaspora for their participation, including the transfer of technology, for India's industrial growth. The establishment of HLCID in 2000 by the Ministry of External Affairs is a landmark in officially recognizing the Indian diaspora that had emigrated right from the early 1830s to the present day, and constructively engaging them for mutual advancement through a number of policy initiatives and programs briefly described above. How these policies and programs have encouraged the states to further facilitate the involvement of diaspora communities in the overall growth of the state and the diaspora is outlined below.

Regional Diasporas and Development

While the major policy initiatives covering the entire Indian diaspora are formulated at the national level by the Government of India, the states have their own mandate with those who had emigrated from its region. State reformation in India took place on the basis of region and language, which often converge. There is an emergence of regional/linguistic diasporas which play a significant role in synergizing for mutual development. Foremost among the regional diasporas are Punjabi, Gujarati and Telugu diasporas and the states of Punjab, Gujarat, Kerala and Andhra Pradesh have been pro-active to facilitate investments and developmental projects by their diasporas. These states have already put in place necessary institutional arrangements to facilitate NRIs and NRI diaspora to take up developmental projects, including philanthropical ones, in locations of their choice along with a number of incentives such as land, power and water connectivity and tax-holiday besides matching grant for starting projects in priority sectors and backward areas.

Several regional diasporas have also formed global diaspora organizations to collectively advance economic, cultural, economic and socio-political interests of the region as well that of their diasporic community dispersed across the globe. For instance, the *Global Malayalee Council* (http://www.globalmalayalee.com/), the largest organization of residents and non-residents of Kerala, has set the goal for the comprehensive progress of India, Kerala in particular, through promoting Community Welfare, Cultural Exchange, Health and Wellness, Business and Trade and Educational advancement. Other such organizations include World Telugu Federation (http://www.worldtelugu.co/),

Vishwa Gujarati Samaj (http://www.vishwagujaratisamajusa.org/), Non-Resident Gujarati Foundation (https://nri.gujarat.gov.in/guj-non-resi-found.htm), World Punjabi Organization (http://worldpunjabi.org/) and Global Punjabi Association (http://www.globalpunjabiassociation.com/) are some of the global organizations that synergize regional diaspora communities world over with respective regions in pursuit of overall development of the regional diaspora communities and their places of origin, bringing them together in the pursuit of their common goals of progress. A brief note on Punjabi diaspora is given below to illustrate the pro-active approach of the state of Punjab in mobilizing Punjabi diaspora for the development of Punjab.

Punjabi Diaspora and Development

The Punjabi immigrants remit a high proportion of their earnings to support their families back home and also to improve the economy of the home state. According to the World Bank report of 2004, Punjab receives the highest remittances from NRPs, amounting to nearly $3 billion (Thandi 1996). Punjabis invested their remittances in '...establishing industries, factories, and buying land and transport companies in most of the major towns in Punjab such as Jalandhar, Ludhiana, Chandigarh and other towns' (Tatla 1999: 63). A major part of their remittances goes for the development of social sector such as hospitals and educational centres in Punjab. Contribution of overseas Sikhs in the field of agriculture, for instance, has successfully launched the 'green revolution' during the 1960s.

Several Canadian Punjabi Philanthropists have made contributions towards institution building, health care and education in Punjab. Significant philanthropic projects have been undertaken by the individuals as well as Canadian Punjabi Associations and collectives. One of the examples of this type of transnational philanthropy is the development of ancestral villages into modern 'model villages', comparable to the West, in terms of infrastructural facilities such as underground water, sanitation, proper sewerage and road systems, healthcare besides computer education. The village Kharoudi in Hoshiarpur district is one such example of a project undertaken by VLIF (Village Life Improvement Foundation), established by two NRPs from this village settled in British Columbia in Canada, Dr. Gurdev Singh Gill and Dr. Raghbir S. Basi. Kharoudi comprising of 150 households comprising 700 people was developed with all the modern civic amenities was inaugurated by Mr. Ujjal Dosanjh, Punjabi-Canadian Premier of British Columbia on 26 December 2000. Another group of philanthropists undertook modernization of their village Bahrampur in Ludhiana, following the Kharoudi model with the help of VLIF, and completed the project in 15 months (Kaur 2011).

Individual philanthropists like Budh Singh from Canada established *Guru Nanak Mission Medical and Educational Trust* to provide medical care and education in his native district Nawanshar. Harvinder Singh Saran, another Canadian Punjabi established Harvest Tennis Academy (HTA) in Ludhiana in 2005 offering professional training for boys and girls to participate in national and international competitions.

Government of Punjab started 'Parvasi Punjabi Divas' since 2002, inviting leading businessmen in Punjab and important persons of Punjabi diaspora from different countries where Punjabis are found in significant numbers. Prominent members of the overseas Punjabi community were honored with 'Punjabi Gaurav Sanman'. Punjabi NRIs as well as diaspora are offered incentives for setting up industrial enterprises on a preferential basis with industrial areas reserved in major cities of Punjab, providing 'Single Window Clearance' for investment proposals. In 2002 a separate *Ministry of NRP Affairs* was established, with a Secretariat at Chandigarh, along with *NRP Helpline* to get feedback from NRPs, especially from Canada, UK and the US. 'Fast Track Revenue Courts' are established for the speedy disposal of revenue disputes involving NRPs.

An innovative policy framed in 2002 by Punjab government is *Mera Pind* Initiative, known as Scheme CD 2.35 that offered matching grant to NRPs and PIOs for funds invested on 'development projects' such as the development of rural infrastructure and investing in ancestral villages. The government grant was further enhanced to 75% of the project in 2008 and the scheme is extended to include such projects in urban areas too. Fast clearance of projects is carried out by Punjab NRI Committee for Development headed by the Chief Secretary of the Government. The government share of 25% the project is released as soon as the investor spends 25% of the budget and the rest when the project is completed. The maintenance of the infrastructure is done by the investor, individual or collective like a trust or NGO, through the Village Development Committees (Kaur 2011).

Punjabi diaspora philanthropy and support in the development of sectors like education and healthcare are significant in changing rural

Punjab in particular. Such initiatives have transformed several villages into 'global villages' and 'cyber villages' with remittances and investments of the diaspora communities.

A review of Punjabi Diaspora Philanthropy suggests that there is a huge potential out there among the Indian diaspora communities to undertake philanthropic projects in different sectors for India's development. The government, both the central as well as the state, needs to put in place pro-active coordinating and facilitating institutional structures that are trustworthy, free of bureaucracy and red tape, to motivate the diaspora.

To Conclude—Some Reflections

As immigrants, Diaspora communities work hard to earn their livelihood and to contribute to the economic progress of their host nations and also that of motherlands. Immigrants by definition need to show their worth by exhibiting their skills and services at lowest wages if they are to be employed by the hosts and create a demand thus carrying out developmental works to the nation-building of their host society at the lowest cost possible. Globalization at the dawn of twenty-first century has been propelling ***diasporaization*** (Cohen 1999) hitherto never envisaged. Migration and immigration beyond borders, however restrained and challenged by nation-states, are beyond control (Bhagwati 2003) and only add to the emerging power of the diasporas. The challenge before diasporas today is to organize themselves to grab the emerging opportunities for socio-economic and political advantages of contemporary globalization characterized by borderless markets, information highways, global institutions etc., to reconnect with the motherland/ancestral land, for assuming significant role in global affairs.

Today in the highly globalized world diasporas are so intertwined with their homelands or ancestor lands that they share similar challenges as much as opportunities for their advancement. While in a few instances, such as the Jewish and the Chinese diaspora, there are already certain institutional and organizational arrangements in place between diaspora and the 'motherland', India initiated dialogue with her diaspora only with the dawn of twenty-first century. Globalization today offers great openings for socio-political and economic and advancement if only the diaspora together with the 'motherland' can seize them, in the fields of business, commerce and industry, technology, education, healthcare, culture, etc., including international relations in this highly interdependent world. The geo-political and economic interests of the new nation-states are safeguarded and promoted if their diasporas are well integrated and dispersed worldwide.

A major element in the advancement of industrially developed countries has been the immense contribution of people and resources from developing and underdeveloped world, both under imperial and neo-imperial contexts. The bridging of diasporas with motherland or ancestor-land is progressively reversing the flow of wealth and social capital to the roots. The pro-active policies of the developing countries to partner with their diasporas is increasingly augmenting this process. The diasporas too are demanding greater participation in the affairs of the nation-state, both within as well as at the international arena. For instance, the two branches of Global Organization of People of Indian Origin, having their head offices in New York and Mauritius respectively, have been holding Annual Conventions during the Pravasi Bharatiya Divas celebrations right from its inception in 2003. These regular Annual Conventions offered the Diaspora delegates' opportunities to interact with their counterparts in India and also the Indian leaders from economic, political and socio-cultural domains for furthering the inter-linkages and partnerships. 'Overseas Citizen of India' is one such outcome to the demand for dual citizenship by NRI Diaspora, articulated through the GOPIO, which offers diaspora all entitlements of an Indian citizen except participation in Indian elections—both as a voter and contestant—and also in the Indian Civil Services. But GOPIO is still hopeful and has resolved seeking full citizenship to be conceded if they are to meaningfully partner and invest in India's growth story to the fullest extent, as it is prevalent in many countries where dual citizenship granted.

India today, as mentioned earlier, exists stretching far beyond her borders sharing equally the happiness on her achievements in Indian Space Research Organization as much as Indian Silicon Valley inventions in California. Needless to say that any disaster or tragedy striking any Indian anywhere in the world is a matter of concern to all Indians wherever they are. The news of 25 September 2016 (The Hindu 2016: 11) from Washington that the Indo-Americans have launched a White House petition requesting the President of USA to designate Pakistan 'a state sponsor of terrorism', following a Bill to this effect was tabled in the US Congress by two influential Congressmen following the Pakistani terrorist attach killing 18 Indian soldiers at Uri in Jammu and Kashmir, amply illustrates emerging India beyond India.

References

Appadurai, Arjun. 1990. *Modernity at Large: Cultural Dimension of Globalization*. Minneapolis: University of Minnesota Press.

Bhagwati, Jagdish N. 2003. Borders Beyond Control—A Door that will not close, *Foreign Affairs*, January/February 2003 Issue, p. 98–104.

Bhat, Chandrashekhar. 2003. 'India and the Indian Diaspora: Inter-linkages and Expectations', in Ajay Dubey (ed.), *Indian Diaspora: Global Identity*, pp.11–22. New Delhi: Kalinga Publications.

Bhat, Chandrashekhar and T.L.S. Bhaskar. 2011. 'Locality and Identity in the Indian Diaspora: Reinvention of Regional / Linguistic Diversities in Mauritius' in N. Jayaram (ed.), *Diversities in the Indian Diaspora*, pp.114–124, New Delhi: Oxford University Press.

Bhat, Chandrashekhar. 2006. Unpublished *keynote* address on "Challenges of Diaspora in the Globalizing World" at the *Armenian Diaspora Forum III Conference*, Government of Armenia, Yerevan, September 18–20.

Bhat, Chandrashekhar. 2012. "Indian Diaspora and Global Organizations: Communities and Contested Boundaries", presented at the International Conference on *India beyond India*, Gottingen University.

Cheran, R. 2004. "Diaspora Circulation and Transnationalism as agents for change in the Post Conflict Zones of Sri Lanka",A policy paper submitted to the Berghof Foundation for Conflict Management, Berlin. http://www.sangam.org/articles/view2/523.pdf.

Clarke, Colin, Ceri Peach and Steven Vertovec (eds.). 1990. *South Asians Overseas: Migration and Ethnicity*. Cambridge: Cambridge University Press.

Cohen, Robin. 2008. *Global Diasporas: An introduction*, Second edition, New York: Routledge.

DellaPergola, Sergio. 2013. *World Jewish Population*, Jerusalem: The Avraham Harman Institute of Contemporary Jewry, The Hebrew University of Jerusalem. http://www.jewishdatabank.org/studies/downloadFile.cfm?FileID=3113.

Eisenstadt, Michael *and* David Pollock. 2012. "Friends with Benefits: Why the U.S.-Israeli Alliance Is Good for America,", *Foreign Affairs*, November 7. https://www.foreignaffairs.com/articles/2012-11-07/friendsbenefits?gp=135637%3A36ce918050c21605.

Hermann, Elfriede and Roman Loimeier. 2012. *India beyond India: Debating Communalism and Belonging*, May 24–26, University of Gottingen.

Kapur, Devesh and John McHale. 2005. *Give Us Your Best and Brightest*, Washington: Centre for Global Development.

Kaur, Mandeep. 2011. *Canadian-Punjabi Philanthropy and its Impact on Punjab: A Sociological Study*. Unpublished Ph.D. Thesis, Patiala: Punjabi University.

Kuznetsov, Yevgeny, and Charles Sabel. Forthcoming. "Work Globally, Develop Locally: Diaspora Networks as Springboards of Knowledge-Based Development." *Innovation: Management, Policy, and Practice* 8 (1–2). Washington: WBI Deveolpment Studies.

Kuznetsov, Yevgeny and Charles Sabel. 2006. "International Migration of Talent, Diaspora Networks, and Development: Overview of Main Issues" in Kuznetsov, Yevgeny (Ed.) *Diaspora Networks and the International Migration of Skills: How Countries can Draw on their Talent Abroad*. Washington: World Bank Institute Development Studies. pp. 3–18.

Lechner, Frank, *Globalization*. In: George Ritzer (ed.), Encyclopedia of Social Theory. Vol I., Thousand Oaks—London—New Delhi 2005, pp. 330–330.

Levitt, Peggy. 1999. *Towards an understanding of Transnational community: Forms and their Impact on Immigrant Incorporation*. Harvard University: paper presented at 'Comparative Immigration and Integration Program', Winter Workshop University of California at San Diego, 19 February.

Manaseryan, Tatoul 2004 *Diaspora the Comparative Advantage for Armenia*, Working Paper No. 04/14, Armenian International Policy Research Group, downloaded from http://edoc.bibliothek.uni-halle.de/servlets/MCRFileNodeServlet/HALCoRe_derivate_00002974/Diaspora.pdf.

Ministry of External Affairs. 2002. *Report of the High Level Committee on Indian Diaspora* (HLC). Ministry of External Affairs, Government of India, New Delhi. (Also at http://indiandiaspora.nic.in/contents.htm).

N. Jayaram (ed.). 2011. *Diversities in the Indian Diaspora*. New Delhi: Oxford University Press.

Nonini, D.M. 2000. "Diasporas and Globalization". In: Malvin Ember, Carol R Ember, Ian Skoggard (Eds.): *Encyclopedia of Diasporas*. New York: Springer Science+Business Media Inc., pp. 565–572.

Robertson, Roland. 1992. *Globalization: Social Theory and Global Culture*. London: Sage Publications.

Safran, William.2012. *Israel and the Diaspora Problems of Cognitive Dissonance*. Working Paper 53 (April) Oxford: International Migration Institute.

Seagrave, Sterling (1995) *Lords of the rim: the invisible empire of the overseas Chinese*, New York: G. P. Putnam & Sons

Tatla, Darshan Singh. 1999. *The Sikh Diaspora: The Search for Statehood*. UK: UCL Press Ltd.

Thandi, Shinder. 1996. 'The Punjabi Diaspora in the UK and the Punjab Crisis', in Pashaura Singh & N. Gerald Barrier (eds.), *The Transmission of Sikh Heritage in the Diaspora*. New Delhi: Manohar.

Thandi, Shinder S., "Punjabi Diaspora and Homeland Relations" http://www.india-seminar.com/2006/567/567_shinder_s_thandi.htm, downloaded on 18/2/2013. New Delhi: Manohar Publishers.

The Hindu. 2016. "Indian-Americans to Petition Obama", 25 September, p.11.

Tinker, Hugh. 1974. *A New System of Slavery: The Export of Indian Labour Overseas 1830-1920.* London: Oxford University Press.

UK Foreign and Commonwealth Office. 2015. *Armenia's Diaspora—Its Role & Influence,* https://www.gov.uk/government/uploads/system/uploads/.Armenia_II_rev_1.doc.

Vertovec, Steven and Robin Cohen (eds.). 1999. *Migration, Diasporas and Transnationalism.* Cheltenham, UK: Edward Elgar Publishing Ltd.

Zachariah KC, Irudaya Rajan S 2004. Gulf Revisited. Working Paper 363, Thiruvananthapuram: Centre for Development Studies.

Zachariah K C, Irudaya Rajan S. 2011. *Inflexion in Kerala's Gulf Connection: Report on Kerala Migration Survey 2011.* Working Paper 450. Thiruvananthapuram: Centre for Development Studies. (http://www.cds.edu/wp-content/uploads/2012/11/WP450.pdf).

The Diaspora Engagement: Role of Fiji-Indian Diaspora in Development of Fiji

Manoranjan Mohanty

Introduction

Diaspora engagement discourse is an important part of migration-development debate. Diaspora engagement has emerged as a development-policy, strategic-support and international relation issue. The role of diaspora in the development of both countries of origin and destination draws greater attention today than ever before. Apart from maintaining geo-ethnic group identity in host country, diaspora brings in remittance, investment, human and social capital, and provides a strategic support to home land. The onset of globalisation has stimulated the diasporic movements and social and economic networking and in turn, greater diasporic engagement. Cheaper means of communication and travel, and growth of mass media and ICT have contributed much to the movement of people across border, creating 'transnational communities', globally. The relationships between migration, diaspora and development need to be fully understood.

M. Mohanty (✉)
The University of the South Pacific, Suva, Fiji
e-mail: manoranjan.mohanty@usp.ac.fj

A. Dubey and S. Rajwar (eds.), *Diaspora in Development*,
https://doi.org/10.1007/978-981-96-6747-5_3

Indian diaspora is one of the largest Diasporas in the world. Indian Diaspora (including PIOs/ OCIs and NRIs) in the Pacific especially Fiji, constitute a substantial number, rooted to the British indenture labour system. The indenture labourer called '*girmitiyas*' that came to Fiji under "girmit" (Agreement) between 1879 and 1916 and the free migrants including Punjabis, Gujaratis that came later settled in Fiji., they formed a substantial Indian diasporic communities and contributed significantly to Fiji's economy and society and to the socio-cultural development, making Fiji, a modern nation-state. Despite wide linguistic and religious diversities that exist among Fiji-Indian Diasporic communities, the group identity is however preserved in Fiji.

The paper brings out theoretical linkages between diaspora and development. It provides aspects of Fiji-Indian diasporic communities' engagement in Fiji from historical and contemporary perspectives, focusing on the role of Fiji-Indian Diaspora in the development of Fiji. It also explores the extent of their engagement to the original homeland.

Migration, Diaspora and Development Linkages

Migration and diaspora are the defining trends of World today. Human migration/mobility is closely linked to development. Generally speaking, the greater the development, more is the mobility of people and vice versa. Migration through diaspora establishes relationships between people and place. Migration is regarded as 'binary mobility' (Sinatti and Horst 2014) between the country of origin and residence.

Diasporas are the 'communities rooted in a national home and sharing a group identity' (Sinatti and Horst 2014). Diasporas are defined broadly as 'transnational communities of a particular kind, characterised by having experienced movement from an original homeland' (ibid.). Today, diasporas are considered as agents of change and new actors in the development of the country of origin and country of residence. Within the diaspora, a strong ethnic group consciousness does exist. There has been a growing recognition that Diasporas have an important place in the economic and political collaboration between countries of residence and of origin (ibid.). Diaspora groups are defined in relation to a home country (a place) and in terms of ethnic group consciousness (people) (ibid.). People, place and identity are the cornerstones of diaspora discourse.

There exists a linear relationship between migration, diaspora and development as follows:

> Migration → Diaspora formation → Diaspora engagement → Development.

Much of diaspora engagement depend upon the processes of diaspora formation. Development is a concept that expresses positive social transformation. Development is generally linked to people and place. However, development is no longer seen as something that happens in a geographically defined space, but rather for the 'people' concerned (Sinatti and Horst 2014: 15). Development today, is 'about creating conditions for people rather than for places' (ibid.).

In the migration-development lexicon, diasporas are 'groups of people originating from a given nation-state' (Sinatti and Horst 2014). They are the group of skilled migrants that 'possess the potential to provide benefits to their countries of origin' (Siddiqui and Tejada 2014). Diaspora is emerging as a social, economic and political concept. The traditional 'brain drain' concept is now replaced by the diaspora discourse. With this discourse, the 'brain drain' approach that emphasised skilled migration as a permanent loss from the country of origin has now changed to the concept of 'brain gain' when the skilled migrants return to home country from abroad (Siddiqui and Tejada 2014; Chacko 2007). Meyer (2001) observed that since the early 1990s, Diasporas have created 'multiple associations and long-distance connections that could benefit the country of origin'. Diasporas give rise to the formation of 'transnational communities' (Lowell and Gerova 2004; De Haas 2006; Tejada 2012 cited in Siddiqui and Tejada 2014). The diaspora concept views skilled migrants as carriers of a 'social capital' and this capital needs to be organised and harnessed for the benefit of the home country (ibid.).

Various empirical studies (Agunias and Newland 2012; Kuznetsov 2013; CODEV-EPFL et al. 2013) 'illustrate diverse ways in which countries of origin can benefit from skilled migration through both the diaspora and returnees' (cited in Siddiqui and Tejada 2014). Dubey (2011) in his publication provides insights into the Indian Diaspora's contributions to the economies of their adopted homes. Diasporas can bring financial, social and human capital to the development process and the diaspora investment and remittances directly affect economic growth and development. Diasporas contribute significantly to development and nation–building. For instance, development in the West Asian countries e.g., Kuwait, UAE and Oman happened largely due to South Asian immigrants especially by large-scale Indian diaspora communities. The secondary migration from a country of residence to another country and formation of a kind of chain Indian diaspora is a rapid ongoing process, thereby prompting a rapid transnationalisation of Indian diaspora. Migration, diaspora and development nexus has however, received little global attention and the relationships between diaspora and development need to be fully understood. As Sinatti and Horst (2014) point out that the diaspora engagement in development is less understood.

Diaspora Engagement Discourse

Diaspora has been recognised as a new resource and an agent of change and development. Diaspora engagement, therefore, has emerged as development strategies. Diaspora communities have been a major source of remittance, investment, and they also help in improving international trade and relations. Besides, diaspora community provides humanitarian assistance during emergency natural disasters. The diaspora communities have diverse skills and resources that can leverage to promote change in their country of origin (Sinatti and Horst 2014).

The 'migration-development nexus' (Faist and Fauser 2011) is critical in understanding diaspora engagement. Diaspora engagement depends upon the process of migration and diaspora formation. Much of diaspora formation and engagement depend upon the type of migration (forced or voluntary); nature of migration (permanent or temporary); migrants' travel experience or 'routes'; place of origin (peaceful or conflict zone) and so forth. Moreover, 'nationality or ethnicity is a defining feature of the relationship between diaspora and the place of origin' (Sinatti and Horst 2014). In the case of forced and permanent migration, and migrants originating from conflict/political instability areas, ethnicity or ethnic identity may define the relationships between people and place and the nature of diaspora engagement. On the other hand, in a situation of voluntary and temporary migration, nationality comes to the forefront in establishing diaspora relationships and engagement. The 'conflation between place and belonging is implicit in diaspora engagement and practice' (Sinatti and Horst 2014).

Today, it is an inter-dependent and 'borderless' world and an era of networking. The onset of globalisation has stimulated the diasporic movements, networking and in turn, greater diasporic engagement. There has been shrinking space and time resulting in transnational community space for interaction and engagement in a 'global village'.

According to Sinatti and Horst (2014) much of the diaspora discourse 'identifies the loyalties and obligations that migrants hold towards their countries of origin as the motivation justifying their engagement'. Diasporas are often engage at the 'local level, usually in their place of origin, where they are familiar with the context, in many cases, have family ties' (The Guardian 2013). The diaspora community engagement also depends upon the bilateral relations between the country of origin and country of residence. The diaspora engagement varies widely from local, national to transnational, and the form of engagement varies from charity, philanthropy, relief and welfare assistance to development.

Broadly, there can be three levels of diaspora engagement (a) Engagement with country of origin

(b) Engagement with country of residence/settlement and (c) Transnational engagements.

The importance of diaspora in development through their effective engagement is widely acknowledged by the policy makers. Many governments in Asia have institutionalised their engagement with their diaspora, for instance, India (Ministry for Overseas Indian Affairs), China (Overseas Chinese Affairs Office), the Philippines (Philippines Commission on Filipinos Overseas), Bangladesh, Sri Lanka and South Korea (The Guardian 2013). The diaspora-engagement organisations go beyond the realm of national governments and to the local and NGO level. Agunias and Newland (2012) noted that the governments at different level-state/provincial or municipal level establish partnerships with diaspora populations to assist in development of the countries of origin'. In India, the governments in some states, such as e.g., Gujarat, Kerala, Odisha, Punjab and Bihar, are actively pursuing diaspora mobilisation by establishing Diaspora offices or Diaspora associations. For instance, the Government of Kerala established the Department of the Non-Resident Keralites Affairs (NORKA) in December 1996 to redress the NRKs' problem. Government of Bihar has been actively engaging with its diaspora through the Bihar Foundation (Times of India, 5 February 2015). Similarly, the Non-Resident Oriya Facilitation Center (NROFC) in Odisha works in collaboration with the Government of Odisha. Various Governments and NGOs are also collaborating together in Diaspora engagement. For example, the Government of South Korea has 'established the Overseas Koreans Foundation, a non-profit organisation affiliated with the Ministry of Foreign Affairs and Trade' (Agunias and Newland 2012).

Global Dimension of Indian Diaspora

The 'transnational Indians' spread world-wide (see Lal 2006; Sahoo and Lal 2009; Lal 2016). India today has nearly 30 million people of Indian origin (PIOs) and expatriate Indians (NRIs) living outside India (The Times of India 2017). The term 'Overseas Citizenship of India' (OCIs) is commonly used today. With globalisation and internationalisation of trade, capital, and labour, the proportion of highly-skilled Indian migrants has increased considerably. Over the last two decades, people of Indian origin represent a highly-skilled—doctors, scientists, engineers, academics, finance professionals, and entrepreneurs—in every country they live in (Greater pacificcapital.com 2013).

Map 1 shows that in 2013, there were 6 million Indian diaspora in the Middle East, 3 million each in USA and Southeast Asian countries, 1.5 million in the United Kingdom, 1 million each in Canada, West Indies and South Africa.

By the early twentieth century, under the British indenture labour system, large-scale Indian migrants were dispersed world-wide. The system of indenture that continued between 1834 and 1916 was responsible for the migration of over one million Indians to work in plantations

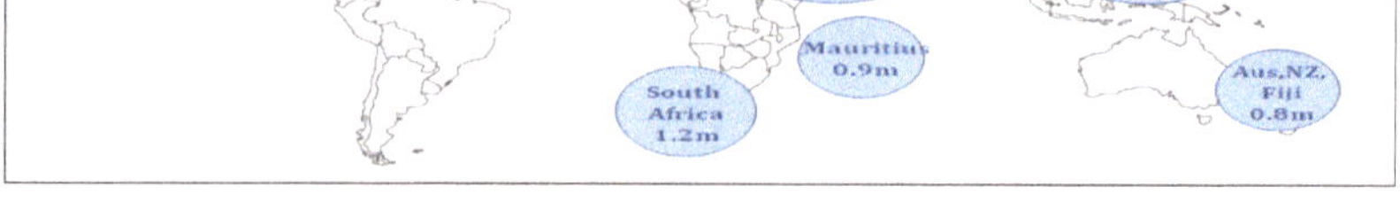

Map. 1 Global Distribution of Indian Diaspora. 2013 *Source* Adapted from Greater pacificcapital.com ()

Table 1 Distribution of overseas Indian diaspora concentration (> 10% of country's population)

Country	*Indian diaspora population (PIO & NRI)*	*% of Country's population*
1.Mauritius	994,500	73.0
2.UAE	3,500,000	57.1
3.Guyana	327,000	44.3
4.Trinidad Tobago	470,000	38.8
5.Fiji	340,000	37.0
6.Qatar	650,000	28.1
7.Suriname	148,000	25.0
8.Kuwait	700,000	24.3
9.St.Vincent & Grenadines	21,500	21.1
10.Saudi Arabia	4,100,000	14.4
11.Nepal	4,000,000	13.6
12.Grenada	12,000	10.7
13.Oman	450,000	10.0

Source Mohanty, M. based on data from Wikipedia (2018) and CIA World Factbook (2018)

in Asia, Africa and Latin America. During 80 years of existence of the system, large-scale Indian migrants were dispersed to various countries, forming a distinct Indian Diaspora abroad (Table 1).

Table 1 shows, the largest proportion of Indian diaspora (PIOs and NRIs) to the country's population are found in Mauritius (73%) followed by UAE (57%); Guyana (44%); Trinidad Tobago (39%), Fiji (37%), Suriname (25%). There are at least 13 nations with 10% or more of Indian origin population to their total population (Table 1).

The countries of Indian diaspora concentration with less than 10% of their population include Singapore, Malaysia, Seychelles, Sri Lanka, Canada, Jamaica, Maldives, St Lucia, South Africa, New Zealand, United Kingdom, Australia, Myanmar, Brunei and USA and others (Wikipedia 2018).

Indian Student Migration

India sends the second largest number of students abroad after China. According to the UNESCO Institute for Statistics, the number of Indian students abroad tripled from about 51,000 in 1999 to over 153,000 in 2007 (cited in Naujoks 2009). Five countries accounted for 90% of

Table 2 Indian Students Diaspora in Selected International Student Receiving Countries

Country	*Indian population 2016/2017*	*% Indian population to country population*	*Indian students 2016/2017*	*% Share of Indian students to total Indian population*
USA	3,456,447	1.0	186,000	5.4
UK	1,451,862*	2.3	16,550	1.1
Canada	1,374,710	4.0	76,000	5.5
Australia	455,389	1.9	70,000	15.4
New Zealand	143,520**	3.6	15,000	10.5
Total	6,881,928	–	363,550	5.3

Source Mohanty, M. based on data from Websites on international students, Wikipedia (2018) and country Census population. * 2011 Census population ** 2013 Census population

all Indian tertiary students abroad. (ibid.) The United States is by far the most important destination country, receiving more than half of the worldwide Indian students (ibid.). In 2016/2017, the Indian student diaspora in USA, Canada, the United Kingdom, Australia and New Zealand was over 360,000, together accounting for nearly 5% of total Indian population in these countries (Table 2). The percentage share of Indian students to the country's Indian population was highest in Australia (15%), followed by New Zealand (11%), Canada (6%) and USA (5%) (Table 2). The United Kingdom received the lowest proportion of Indian students i.e., 1% of its population (ibid.).

Moreover, secondary migration and chain diaspora formation to another country from a country of residence are ongoing processes. When the student diaspora and chain diaspora are added together, the total Indian global diaspora population is much higher than the estimated value.

Indian Diaspora Engagement

Indian diaspora is potential asset for the country's long-term development. While Indian diaspora accounts for only 2.5% of India's population but the total wealth of Indian diaspora community has been 'estimated at US$1 trillion', just half of India's nominal GDP in 2012 (greater pacific capital 2013). Of this, 'half is financial assets, with the rest being land

and physical assets' (ibid.). The income of Indian Diaspora 'community has been estimated approximately US$400 billion' (ibid.). There is thus, a great potential in mobilising Indian diaspora resources for the development of homeland. 'In order to fully capitalize the potential of its diaspora, India clearly needs to find ways to better leverage the overseas Indian community for investment, transfer of know-how, and R&D' (greater pacific capital 2013).

The diaspora engagement can be financial, social, cultural, economic and political nature. Remittance generated by the migrants is one of the economic elements in constructive engagement with the home country. The Middle East countries which have relatively large migrant workers' transfer, share the largest of inward remittances. These countries 'transferred nearly US$70 billion in remittances in 2012' which was 'equal to 3.6% of India's GDP' (greater pacific capital 2013). Remittance is a critical integrating factor of the overseas Indian community with the homeland.

Indian Diaspora in Oceania and Fiji

There were more than half a million Indian Diaspora (PIOs and NRIs) in the Oceania region, of which nearly 40.9% lived in Fiji (Table 3).

Table 3 Overseas Indian Diaspora in Selected Countries in Oceania, 2001 •

Country	*PIO* (A)*	*Indian citizen (B)*	*Total Indian diaspora(A + B)*	*Total population + 2001*	*% Indian diaspora to country population*
Australia	160,000	30,000	190,000	19,338,000	0.98
Fiji	336,579	250	336,829	823,000	40.90
New Zealand	50,000	5000	55,000	3,808,000	1.44
PNG	–	1000	1000	4,920,000	0.02
Solomon Islands	–	20	20	463,000	0. 00
Vanuatu	–	50	50	202,000	0.02
Oceania	546,579	36,320	582,899	30,915,000	1.89

Source Government of India, ministry of external affairs foreign secretary's office (2001)
\+ United Nations Population Division.2001 Revision
* PIO refers to Person of Indian Origin

Historically, under the British indenture labour system, Indian labourers came to Fiji in 1879 to work in sugarcane plantations on a five-year contract and they were called '*girmitiyas*'. The first ship named the *Leonidas*, carrying 463 Indians arrived in Fiji on May 14, 1879. The indenture labour migration process lasted for about 37 years and altogether 87 shiploads of Indians came to Fiji between 1879–1916, with a total number of over 60,000 Indian migrants who arrived in Fiji. The system of indentured labour officially ended on January 1, 1920. Nearly 25,000 returned to India upon the end of the '*Girmit*' (Agreement) and others stayed on to work as farmers, labourers and as petty business people (Naidu and Reddy 2002). Thereafter, Indian immigrants to Fiji came in distinct waves as 'free migrants'. The Indian Diaspora in Oceania (PIOs and NRIs) constitutes nearly 2% of its total population (Table 3) and little less than 2% of the 30 million global Indian Diaspora population.

Figure 1 shows trends in Person of Indian Origin (PIO) in Fiji, since 1881, two years after the first arrival of Indian labour migrants in Fiji under the Indenture labour system. There was a gradual increase in PIOs due to labour migration until the end of migration in 1916. The 1921 Census recorded 60,634 persons Indian Origin and this population subsequently increased to more than half of Fiji's total population in 1966 (Fiji Bureau of Statistics 2008a). The year 1966 can be considered as a year of great 'Demographic Divide' in the history of Indian Origin population in Fiji. Thereafter, the proportion of PIOs started declining gradually.

The annual growth rate of Indian population in Fiji declined to -0.3% in 1986 and the main reason for such a decline was decrease in the fertility rate of Indian population (Fiji Bureau of Statistics 2008b). After 1986, especially after the 1987 coup, the emigration became more intense, and the political instability became a far more important contributor to emigration thereby decreasing the Indian population growth rate in Fiji (Mohanty 2001). The following decades witnessed faster decline in population of Indian origin and the decadal change of Indian population was -2.8 and -7.4 in 1996 and 2007 respectively (Fiji Bureau of Statistics 2008a). In almost five decades, the proportion of population of Indian origin to Fiji's total population has declined from nearly 51% in 1966 to 31% in 2016, and it is projected that, the population of Indian origin in Fiji will further decline to about 26% by 2030 (Fiji Bureau of Statistics 2016).

Figure 1 shows that the share of PIOs in Fiji to the country's total population reached its peak in 1966 and then declined rapidly.

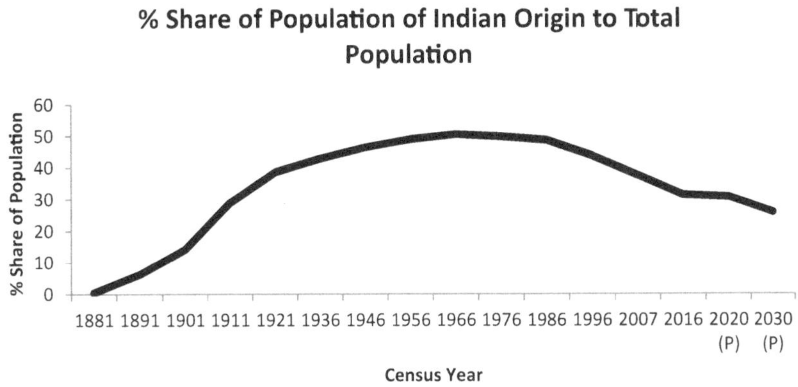

Fig. 1 Trends in Persons of Indian Origin in Fiji, 1881 to 2030. *Source* Mohanty. M based on data from Fiji Bureau of Statistics, 2016.

Indian Immigration and Diaspora Formation in Fiji

The Indian immigration and diaspora formation in Fiji may be broadly categorized into three distinct phases:

The first wave of Indians to Fiji was between 1879 and 1916. During this phase, altogether, over 260 social groups comprising Hindu, castes/ sub-castes, Muslim and tribal groups came to Fiji with a diversified background in terms of religion, caste and region (Lal 1983). Majority of the immigrants was originated from the northern provinces of Uttar Pradesh and Bihar and central India (Ali and Crocombe 1981). In 1888, an unusually high number of single women with families came to Fiji, thirty-four women came with 51 young children and unaccompanied by husbands (Shameem 1990: 150). The '*girmitiyas*' gradually transformed into a distinct Diaspora of Fijians of Indian descent within Fiji and abroad.

The second stream of Indian immigrants to Fiji and elsewhere in Pacific was the influx of a heterogeneous collection of some 2000 to 3000 Indians who came to Fiji either as 'free migrants' or under contracts without penal sanctions (Willson et al. 1990: 102). By 1920, there was a 'sizeable free Indian population' and they were mostly 'farmers from Punjab and traders, jewellers and merchants from Gujarat' (ibid.). The Gujarati immigrants, mostly Hindus first arrived in 1906 (Miller 2008: 64). The community formed a distinct Gujarati diaspora in Fiji numbered about 324 in 1921 and the number increased to 2500 by 1936 (Ali and

Crocombe 1981). The establishment of trade, commerce and business in Fiji started with the arrival of Gujarati community and they 'established local business networks with ties to India' (Miller 2008).

Immigrants from Northern provinces of Punjab especially from the districts of Jalandhar and Hoshiarpur generally sought agricultural works (Gillion 1962: 133 cited in Miller 2008: 64). The earliest Punjabi immigrants arrived in 1904, but the majority of immigrants arrived after 1920 (Miller 2008: 64). The Sikh migrants as business entrepreneurs though form distinct Indian Diaspora elsewhere in the world, they are however, very small in number in Fiji. According to one estimate, the Sikh Diaspora in Fiji consisted of a population between 1200 and 2500 during 1890–1910 (Tatla 1993: 68).

The third and the contemporary wave which is an ongoing process and accelerated since the independence in 1970, is the stream of 'free' and unassisted Indian immigrants. Besides, the traders and businessmen, they include more skilled labour including professionals and technicians that are working in government, private, international agencies and academic institutions. They came either directly from India or from the Pacific Rim metropolitan countries such as Australia, Canada, New Zealand and USA and elsewhere.

Fiji-Indian Diaspora Engagement

The three distinct Indian sub-diaspora groups that exist in Fiji include Girmitiyas and their descendants, Gujarati community and the contemporary free expatriate Indian migrants. These Indian diaspora sub-groups have different level of engagement with the original homeland that needs to be analysed separately.

Girmitiyas and their descendants have limited social and financial engagement with original homeland and their engagement is mainly confined to the country of settlement (Fiji) and transnationally. Today, the Fijian-Indians have formed distinct transnational diaspora abroad through 'secondary' and 'tertiary' migration especially in Pacific-Rim metropolitan countries such as Australia, Canada, New Zealand and USA (Mohanty 2006). The transnational Fijian-Indian diaspora maintained close social networks with Fiji and make occasional visit to their home country-Fiji.

Unlike '*girmitiyas*' and Indo-Fijian diaspora, the Gujarati community in Fiji has maintained personal and business ties with India, remitted capital, and often travel to homeland to get married, and return to Fiji with their brides (Miller 2008).

The third sub-Indian diaspora group i.e., the contemporary 'free' Indian migrants, largely Indian expatriates engage actively with their homeland in terms of social and financial engagement including remittance and investment and networking with families and relatives in India.

These sub-Indian diaspora groups shared common identities and have varied engagement and relationships with the original homeland (India). The relationships of 'old' girmitiya Indian diaspora in Fiji with the original motherland (India) have been relatively limited.

In the recent years, with the active engagement of Government of India with Fiji-based Indian Diaspora, these groups are participating in various educational, social and cultural events in the original homeland. For instance, Fiji was among eight other countries that took part in the first International *Ramayan Mela* in India which was held in New Delhi on 23 February 2015. This was part of cultural engagement and diplomacy in enhancing international relations (The Fiji Sun, 9 March 2015). The Prime Minister of India, Mr Narendra Modi, while inaugurating the Ramayan Mela said, 'India should leverage its great traditions and culture in forging ties with countries around the world in a way that is deeper, more personal, and therefore, far more powerful' (Fiji Sun, 9 March 2015).

In Fiji, the Fiji Seva Sangh along with the Ministry of Education, Heritage and Arts, National Archives of Fiji and the Indian Council for Cultural Relations (ICCR), organised an International Ramayan Conference on 15th–16th October 2016 in Fiji. Eminent scholars, writers and scientists from India, Trinidad, Mauritius, Australia, Fiji and Malaysia participated and engaged in intellectual exchange on various aspects of Ramayan in this conference.

Role of Fiji-Indian Diaspora in Development of Fiji

During 139 years of existence of '*girmitiyas*' and 112 years since the arrival of Gujarati community in Fiji, the Indian Diaspora made significant contributions in making Fiji a modern and prosperous nation. Within five generations, the indentured Indians have transformed themselves from bonded agricultural labourer to small holding farmers, to artisans, to lower and higher ranking government officials, to entrepreneurs and to skilled professionals. Over the years, Fiji-Indians accumulated wealth and they invested in sugarcane farming, trade and commerce. As their wealth increased they invested not only in growing sugarcane but also in trade, commerce and professional services. About 54% of labour force among persons of Indian origin in Fiji was economically active (Fiji Bureau of Statistics 2008b). The sugar cane farming was predominately in Indian hands and they subsequently established garment, and other manufacturing industries. Traditionally, Indians have leased land from indigenous Fijians for sugarcane farming and the lease terms vary from 10 to 99 years. The Fiji-Indians hold about 90% of total cane leased land in Fiji (Naidu and Reddy 2002).

Fiji has been transformed from predominantly agricultural and subsistence economy to a middle income and cash economy. Fiji's trade and commerce are managed largely by Fiji-Indians. Fiji-Indian Diaspora has contributed immensely to almost all sectors of economy including agriculture, business and commerce, trade, transport and communication, industry and to the overall economic development of the country.

Table 4 shows that nearly 39% Fiji-Indians are engaged in salaried occupations. Nearly 54% of Fiji's legislators, senior officials and managers, and 42% of country's professionals and 65% of plant and machine operators and Assemblers, are Indians. Fiji-Indians also occupy almost half of country's technicians and clerical positions. Nearly 41% of Fiji-Indian labour force is skilled agriculture and forestry, and crafts & related trade workers (Table 4).

A bulk of Fijian of Indian descent population (56%) live in urban areas in Fiji (Fiji Bureau of Statistics 2008b). Fiji Indian diaspora contributed significantly to social development in the field of education, health and to the cultural advancement. Fiji-Indians actively participated in politics in the 1980s and 1990s. Fiji witnessed a fast social, cultural, economic and political development as a result of the hard labour and entrepreneurships of Fiji-Indian Diaspora. While the economy is dominated by Gujarati businessmen, the 'girmitiyas' and their descendants predominated the government and social sectors and have contributed to all sectors through their technical and professional skills. Today, the contemporary Indian 'free migrants', accounting a substantial proportion of total immigrants in Fiji, are engaged in government and private sectors and contributing much to social, cultural and economic development of Fiji.

Rural ex-*girmitiya* settlements developed social and political organisations ranging from '*harvest gang*' to school 'committees' (Miller 2008)

Table 4 Occupational engagement of Fiji-Indian population

Occupation	*No*	*% Fiji-Indians in country's occupation*	*% Share of Fiji-Indian engagement to total occupation*
1.Armed forces	73	2.7	0.05
2.Legislators, senior official & managers	9969	54.1	7.5
3.Professionals	11,657	41.8	8.8
4.Technicians & associate professional	10,947	49.8	8.3
5.Clerks	9396	49.9	7.1
6.Service workers & shops & market assistants	9687	30.6	7.3
7.Skilled agriculture & forestry workers	33,981	26.0	25.6
8.Crafts & related trade workers	20,349	56.4	15.4
9.Plant & machine operators & assemblers	14,063	64.7	10.6
10.Elementary occupation	12,433	36.6	9.4
Grand total	132,555	38.6	100.0

Source Mohanty, M. based on data from Fiji Bureau of Statistics (2016)

and 'farmer organisations'. Subsequently, various Indian diasporic organisations have been emerged in Fiji that have contributed significantly to welfare and social and cultural development especially in the field of education and human resource, and art, music, and dance. Notable among them are e.g., Arya Pratinidhi Sabha, Arya Samaj, Dakhina India Andhra Sangam, Fiji Seva Sangha, Gujarat Samaj, Gujarat Education Society of Fiji, Kshatriya Society, Muslim League, Sanatan Dharma Pratinidhi, Sanatan Dharma Sabha, Sathya Sai Service Organisation and Sikh associations of Fiji.

Educational Development in Fiji: Role of Fiji-Indian Organisations

The primary and secondary education sector is uniquely placed in Fiji in the sense that of the 928 schools in Fiji in 2014, there were about 98.5% schools run by NGOs and private sector organisations and only 1.5% controlled by the Government agency i.e., Ministry of Education (Table 5). About 12% schools excluding school committees are controlled and run by Fiji-Indian organisations.

Among the Indian organisations, Arya Pratinidhi Sabha, Sanatan Dharam Pratinidhi Sabha, Fiji Muslim League, Then India Sanmarga Ikya (TISI) Sangam, Gujarat Education Society and Dakshina India Andhra Sangam account for nearly 85% of schools that are run by Fiji-Indian organisations (Table 6).

The Arya Samaj was the first religious, cultural and educational Indian organisation which was established in 1904 in Fiji (Wikipedia 2018). The first school built by Arya Samaj in Fiji in 1918 was the Gurukul Primary School in Saweni, Lautoka. From its inception, Arya Samaj attracted the young, educated and progressive Hindus into its fold (ibid.). The Arya Samaj continued its educational activities and built numerous educational institutions including the University of Fiji.

Table 5 Schools in Fiji by controlling organisations, 2002–2014

Controlling authority	*No of schools*		*% Share 2014*	*% Change 2002–2014*
	2002	*2014*		
I.NGO/ private Sector	916	914	98.5	−0.2
Indian Organisations	129	114	12.3	−11.6
Other organisations	787	800	86.2	−1.7
II. Govt. Sector				
Ministry of education	21	14	1.5	−33.3
Grant total	937	928	100.0	−1.0

Source Mohanty, M. based on data from Fiji Ministry of education, national heritage, culture and arts (2014)

Table 6 Distribution of schools controlled by Indian organisations in Fiji

Indian Organisation	*Number of schools, 2014*
1. Arya Pratinidhi Sabha	23
2. Sanatan Dharam Pratinidhi Sabha	16
3. Fiji Muslim League	22
4. TISI Sangam	21
5. Macuta Muslim League	8
6. Gujarat Education Society	8
7. Dakshina India Andhra Sangam	7
8. Sikh Community	4
9. Ahmadiya Muslim Association	4
10. Sathya Sai Service Organisation	1
Total	114*

Source Fiji Ministry of education, national heritage, culture and arts (2014)
* This excludes other private schools

Art, Music and Cultural Development in Fiji

Fiji became a multicultural nation with a rich cultural tradition and diversity. Fiji Indians came from a variety of sub-cultures and religious groups, but as a group they have retained many of their cultural and religious traditions, including music, dance, festivals and cultural values.

About 77% of Indian origin population in Fiji were Hindus in 1996, the dominant sect being Sanatan Dharm that accounted for about 74% of the Hindu population followed by Arya Samaj of 3.6% (Fiji Bureau of Statistics 2008a). Other minor Hindu sects include Kabir Panthi and Satya Sai Baba. The other religious faiths among Fiji Indian Diaspora community include Islam, Christianity and Sikhism.

Local folk music including tambura bhajan, qawali kirtan, lok geet, ramayan katha, Ramlila, and chautal is popular among the Hindu diaspora (Miller 2008: 181). The 'folklore of the Fijians of Indian descent derives from traditional Indian folklore' and the epic stories and myths such as the *Rāmāyana* and the *Mahābhārata* are wide-spread and these are read, and chanted by Indians in Fiji at ceremonies and celebrations (ibid.).

Role of Ramayan Mandalis in Fiji

Ramayan Mandalis play a critical role in diaspora engagement. The *Ramayan Mandali* emerged during the 1930s and 1940s as the fundamental institution of Sanatan Dharam practice in Fiji, partly through the influence of visiting Hindu missionaries from India (Miller 2008: 252). Ian K. Samarville observed that three out of every ten adult Sanatani males were members of Ramayan Mandali with an estimated number of 400 to 500 Ramayan Mandalis Fiji-wide (cited in Miller 2008). The Fiji's High Commissioner to India at the Ramayan Mela inauguration in new Delhi said, *Ramayana and Ramayana Mandalis* have a constructive role in the social fabrics of Fiji for more than 130 years (Fiji Sun, 9 March 2015). Such groupings known as *'mandalis' which* are a source of inspiration for the youth (ibid.). Every settlement in the country has unique *Ramayana Mandalis* and their recitations on a weekly basis help to educate and preserve religion, cultural and heritage in Fiji (ibid.). Apart from cultural engagement, the *Ramayan Mandalis* provide charity, natural disaster relief assistance and play a critical role in educational, social and cultural development. The *Ramayan Mandalies* in Fiji also maintain their transnational engagement, networks and relationships.

Indian diaspora in Fiji as elsewhere are more 'divergent diaspora'. A high degree diversity exists within Indian diaspora in Fiji in terms of religion, region and language. It can include diversified sub-groups such as Hindu, Muslim and Sikh diaspora; and Gujarati, Tamil, Telugu, Marathi, and Malayalam diaspora.

Emigration and Transnational Fiji Indian Diaspora

The changing political economy in Fiji since independence led to large-scale outflow of Indians mainly of the skilled categories to the Pacific Rim metropolitan countries such as Australia, New Zealand, Canada and USA (Mohanty 2001, 2006). The emigration trend accelerated in the events of the coups and political instabilities in 1987 and 2000, the non-renewal of leases of land since 1997 and resulting in land insecurity (ibid.). Both push (political instability and land insecurity) and pull factors (e.g., greener pastures) prompted increasing Indian emigration from Fiji. The Fiji-Indian diaspora emigration averaged nearly 5000 people a year between 1987 and 1997, most of those leaving were skilled and financially secured individuals (ibid.).

There has been rapid secondary migration of Indian population from Fiji to other countries especially to the Pacific Rim Metropolitan countries such as Australia, New Zealand, Canada, and USA (Mohanty 2006) and formation of a chain Fiji-Indian diaspora. The trans-nationalisation of Fiji-Indian diaspora is a rapid process in the last two decades or so.

Challenges

The emigration of professional Fiji-Indians has led to large-scale 'brain drain' on Fiji's skilled labour force. The land tenure insecurity is one of the major concerns for Fiji Indians. Majority of sugar land leases with Indian farmers began expiring since 1997. However, the Fiji Ministry of Lands has 'removed past restrictions' of 'one agricultural lease per applicant', 'allowing farmers to expand their capacities beyond one lease' and renewed some agricultural leases for sugarcane farmers in 2015 and planned more renewal of 'expired sugarcane leases in the Western and Northern divisions' (The Fiji Times, 20 October 2015).

The decrease in fertility rate among Indian population in Fiji primarily because of large-scale emigration of reproductive group, is another major concern. The unprecedented Indian emigration from Fiji and the declining fertility rate among Indian population, thus resulting in declining of population of Indian origin in Fiji. The rapid decline of Fiji Indian population has far-reaching social, economic and political implications and a challenge to diaspora engagement.

Conclusion

Today, the Diaspora has been recognised as a new resource and an agent of change, and development. The diaspora engagement has emerged as a development-policy, and strategic-support issue. The Indian immigration under British Indenture labour system had germinated seeds for contemporary Indian diaspora formation, globally. In Fiji, the 'old' Indian girmitiyas and their descendants have gradually transformed into distinct Fiji-Indian diaspora today. While the Fiji-Indian girmitiya diaspora have engaged intensely in the home country and contributed significantly to the development of Fiji, their relationships and engagement with original homeland (India) have been relatively limited. However, there exist a great potential for their engagement in development of the homeland. In order to utilise the potential of Indian diaspora generally, and Fiji-Indian diaspora particularly, India needs to engage more with them, and design strategies and policy measures for their active engagement and mobilisation of their resources that can contribute to its development on a sustainable basis.

References

Agunias, D.R. and K. Newland. 2012. *Developing a Roadmap for Engaging Diasporas in Development—A Handbook for Policymakers and Practitioners in Home and Host Countries*; Geneva and Washington D.C.: International Organisation for Migration and Migration Policy Institute.

Ali, A. and R. Crocombe. 1981. *Pacific Indians,* Institute of Pacific Studies, The University of the South Pacific, Suva, Fiji. pp. 7–9.

Chacko, E. 2007. "From brain drain to brain gain: reverse migration to Bangalore and Hyderabad, India's globalizing high tech cities", *GeoJournal,* 68(2): 131–140.

CIA (Central Intelligence Agency) World Factbook, 2018. Country Profile: population. Retrieved from: https://www.cia.gov/library/publications/the-world-factbook/(Accessed on 25 April 2018).

CODEV-EPFL, IDSK, JNU and ILO 2013 *Migration, scientific diasporas and development: Impact of skilled return migration on development in India.* Final research report. Retrieved from:http://infoscience.epfl.ch/record/188059/files/Migration_ScientificDiasporas_Development.pdf (Accessed on 18 October 2017).

De Haas, H. 2006. Engaging Diaspora. How Governments and Development Agencies can Support Diaspora Involvement in the Development of Origin Countries. IMI Report, International Migration Institute, Oxford university, United Kingdom.

Dubey, A. K. ed. 2011. Indian diaspora: contributions to their new home, New Delhi: Organisation for Diaspora Initiatives in collaboration with MD Publications.

Faist, T. and M. Fauser 2011. "The migration-development nexus: toward a transnational perspective". In Faist, T., M. Fauser and P. Kivisto (eds.) *The migration-development nexus,* 1-26. Basingstoke: Palgrave Macmillan.

Fiji Bureau of Statistics 2008a. 2007 Population and Housing, Suva, Fiji, Fiji Bureau of Statistics.

Fiji Bureau of Statistics 2008b. Statistical News: 2007 Census Results: Population Size, Growth, Structure and Distribution, Suva, Fiji, Fiji Bureau of Statistics.

Fiji Bureau of Statistics 2016. 2013–2014 Household Listing Exercise, Suva, Fiji, Fiji Bureau of Statistics.

Fiji Ministry of Education, National Heritage, Culture and Arts 2014. Annual Report, 2014, Suva, Fiji, Fiji Ministry of Education, National Heritage, Culture and Arts.

The Fiji Sun.March 9, 2015. Fiji At First World Ramayana Mela, March 9 2015. Retrieved from: http://fijisun.com.fj/2015/03/09/fiji-at-first-world-ramayana-mela/ (Accessed on 18 September 2017).

The Fiji Times. 20 October 2015. 99 -year leases for 84, Retrieved from: http://fijitimes.com/story.aspx?id=326511 (Accessed on 20 September 2017).

Gillion, K. L. 1962. *Fiji's Indian Migrants: A History to the End of Indenture in 1920*, Melbourne: Oxford University Press.

Government of India, Ministry of External Affairs Foreign Secretary's Office, 2001. L. M. Singhvi Report: Report of the High Level Committee on the Indian Diaspora, Ministry of External Affairs Foreign Secretary's Office, New Delhi.

Greater pacific capital 2013. Map of the Indian Diaspora, Retrieved from: http://greaterpacificcapital.com/march-2013/ (Accessed on 15 October 2017).

The Guardian. 2013. 'How to improve diaspora engagement'. Retrieved from: https://www.theguardian.com/global-development-professionals-network/2013/feb/20/improving-asian-diaspora-insitutions (Accessed on 15 September 2017).

Kuznetsov, Y. ed. 2013.*How talent abroad can induce development at home? Towards a pragmatic diaspora agenda*, Washington, D.C: Migration Policy Institute.

Lal, B. V. 1983. "Girmitiyas: the origins of the Fiji-Indians". *Journal of Pacific History*. Canberra.

Lal, B. V. ed. 2006. *The Encyclopedia of the Indian diaspora*, Singapore: Editions Didier Millet.

Lal, B. V. 2016. Dispersals, migrations, diversity of communities and the notion of an Indian diaspora, London, Routledge.

Lowell, L. and S. G. Gerova 2004 *Diasporas and economic development: State of knowledge*, Washington DC: Georgetown University, Institute for the Study of International Migration.

Meyer, J.B. 2001. "Network approach versus brain drain: lessons from the diaspora", *International Migration*, 39(5): 91–110.

Miller, K. C. 2008. A Community of Sentiment: Indo-Fijian Music and Identity discourse in Fiji and its Diaspora, Ph. D thesis, University of California, Los Angeles.

Mohanty, M. 2001. "Contemporary Emigration from Fiji: some trends and issues in the post-Independence era" in Naidu, V., Vasta, E. and Hawksley, C. (eds.), *Current Trends in South Pacific Migration*, APMRN, Working Paper No.7, University of Wollongong, Wollongong, Australia, 54–73.

Mohanty, M. 2006. "Globalisation, New Labour Migration and Development in Fiji *Islands"*, S. Firth (ed.),*Globalisation, Governance in the Pacific Islands*,, State, Society and Governance in Melanesia Program Studies in State and Society in the Pacific, No. 1, 107–120. Australian National University, Canberra, Australia.

Naidu, V. and M. Reddy. 2002. *Na Ghar Ke na Ghat ke; ALTA and expiring land leases: Fijian farmers' perceptions of their future*, Ford Foundation Project, MOST, Asia-Pacific Migration Research, Centre for Development Studies, The University of the South Pacific, Suva, Fiji.

Sahoo, A. K. and B. V. Lal eds. 2009. Transnational migrations: the Indian Diaspora, New Delhi; London: Routledge.

Shameem, S. 1990. "*Girmitiya* Women in Fiji: Work, Resistance and Survival".In C. Moore, J. Leckie and D. Munro (eds). *Labour in the South Pacific*. 148–154. James Cook University, Northern Queensland, Townsville.

Siddiqui, Z. and G. Tejada, 2014. Development and Highly Skilled Migrants: Perspectives from the Indian Diaspora and Returnees , International Development Policy, Retrieved from: http://poldev.revues.org/1720. (Accessed on 18 September 2017).

Sinatti, G and C. Horst 2014. "Migrants as agents of development: Diaspora engagement discourse and practice in Europe", *Ethnicities,* 0(0): 1–19.

Tatla, D.S. 1993.The Politics of Homeland: a study of ethnic linkages and political mobilisation amongst Sikhs in Britain and North America, Ph. D thesis, Centre for Research in Ethnic Relations, University of Warwick.

Tejada, G. 2012 "Mobility, knowledge and cooperation: Scientific diasporas as agents of development", *Migration and Development,* 10(18): 59–92.

The Times of India.5 February 2015, Bihar Foundation spreads in 9 countries, Retrieved from:http://timesofindia.indiatimes.com/city/patna/Bihar-Foundation-spreads-in-9-countries/articleshow/46125856.cms (Accessed on 18 September 2017).

The Times of India. 9 January 2017. Prime Minister Narendra Modi's Speech at Parvasi Bharatiya Divas, 9 January, 2017, Bangaluru.

Wikipedia. 2018. Non-Resident Indian and Person of Indian Origin. Retrieved from: https://en.wikipedia.org/wiki/Non-resident_Indian_and_person_of_Indian_origin (Accessed on 25 April 2018).

Willson, M., C. Moore, and D. Munro 1990. "Asian workers in the Pacific", in C. Moore, J. Leckie and D. Munro (eds). *Labour in the South Pacific*. James Cook University, Northern Queensland, Townsville.

Key Role of Indian Diaspora in Réunion Island Development

Jean Regis Ramsamy

A Short Overview of Réunion History

Réunion Island, covering an area of 2,512 km, has seen its size gradually expand in recent decades due to volcanic activity. Originally named *Dina Morgabin* by early explorers in the sixteenth century, the island formed an integral part of the Mascarenes Archipelago.

The first official settlers arrived in 1665 from Fort-Dauphin, Madagascar, led by the Frenchman Étienne Regnault and accompanied by several Malagasy individuals. However, Indians had reached Réunion earlier, referring to it as *Bourbon*. Geographically, the island forms a central link in the chain between Madagascar and Mauritius.

Although the Portuguese explorer Pedro de Mascarenhas is credited with the official European discovery of the island, it was African settlers—locally referred to as *Cafres*—who introduced commercial gardening, especially coffee cultivation. By the early nineteenth century, the focus had shifted to sugarcane, seen by plantation owners as the crop of the future. Following the abolition of slavery in 1848, formerly enslaved

J. R. Ramsamy (✉)
Université de La Réunion, Saint-Denis, France
e-mail: jean.regis.ramsamy@wanadoo.fr

A. Dubey and S. Rajwar (eds.), *Diaspora in Development*,
https://doi.org/10.1007/978-981-96-6747-5_4

people refused to continue working for planters under exploitative conditions. As a result, planters turned to India, long recognized as a hub for human and cultural exchange, to meet their labour demands.

The history of Indians on Réunion became deeply intertwined with that of the island itself. More than three centuries later, the Indian community remains a vital part of the population,[1] representing approximately one-third of its inhabitants. Despite this enduring presence, the Tamil community in particular has yet to fully organize itself socially and politically.

The ROI, Réunionese of Indian Origin,[2] have left an indelible mark on the development of this island. Sugarcane was the main reason for the arrival of an intensive Indian labour force. The Indians arrived to work in the sugarcane plantations.

In the near future, we will come back on one of the questions which some historians asked, *"Did the Indians save the economy of Réunion Island?"*. The issue at hands concerns some arguments which reflect the contribution of India in this ancient colony. We take the risk of unwinding a collection of real facts that deserve a longer study.

Girmityas or Malbar?

The Indian community in Réunion Island is composed of multiple distinct groups. Among them are individuals from Pondicherry, of French nationality, who arrived after 1950. They were followed by the "Karanas"—Persons of Indian Origin (P.I.O.) who had lived in Madagascar, in some cases for nearly two centuries, and who were predominantly of the Islamic faith. (WEBER 1988)

Towards the end of the indentureship period, the first Muslim migrants directly from India began to arrive. Locally referred to as "Zarabes" or Indians from Bombay, many of them came after 1870, often stopping over in Mauritius before reaching Réunion. Their numbers were modest, with around 200 individuals recorded in 1877 and 377 by 1907. Initially,

[1] The Indian population represents 30 to 35% of the global population, i.e. 300,000 people. But in France, statistics are not only allowed for study purposes.

[2] Joseph Varondin, a Réunionese of Indian origin, coined the term ROI. A wink of the eye.

to the term *ROI* as king.

most were granted residence permits, and over time, many acquired full citizenship.

Finally, Indians from Réunion primarily refer to the vast majority of the descendants of indentured workers called "Malbar". In the mid-nineteenth century they were asked to contribute at the rise of sugar cane. This ethnic group of Indo-Réunionese is also characterized by its belief in Hinduism and to a lesser extent in the Church. Between the first convoys from 1828 and the end of the waves of indentureship in 1882, more than 120,000 Indians were introduced into the Colony. This study mainly concerns these Indians.

The term Girmitya does not practically appear in the papers about Indian immigration in Réunion. In the French language, the reserved term is the *engagisme*, by extension the Indian worker commits himself. In Shakespeare's language, they use the words *indentured labours*, or the *Coolie-trade* to translate the phenomenon. In New Delhi, during the last Pravasi Bharatiya Diwas (PBD)[3] at the Ministry of External Affairs, in spite of the omnipresence of the word, we had the opportunity to underline this historic point. In the case of Réunion Island, the term *Malbar/ Malabare* benefits from an ancient legitimacy.

An unexpected development emerged from these early encounters: the origin of the term "Malbar" or "Malabare." This designation can be traced to the fact that many of the first Indo-Portuguese women were associated with the trading post of Goa, that young slaves often came from Goa, and that several other early Indian migrants originated from the Malabar Coast, including regions such as Cochin and Goa. As a result, the descendants of Indian indentured workers—regardless of their specific region of origin came to be collectively referred to as "Malbars." Yet it is a fact that besides the powerful Tamil proportion (thus *Malbare*), components from Calcutta, Andhra Pradesh, Bihar, Gujarat, Maharashtra (Bombay), and a few hundred elements from Orissa came to Réunion island. The local administrative officers during the registration of these workers found a certain ease to indicate the *Malabare* name everywhere.

The recruitment of indentured workers primarily took place in South India, in regions such as Karikal, Madras, Pondicherry, and Yanam. This explains the strong presence of individuals from the southern Indian

[3] Every two years, a meeting of PIO (Person of Indian Origin) representatives is held in India, organized by MEA (Ministry of External Affairs) and Global Organisation of People of Indian Origin (GOPIO).

communities, still recognizable today through surnames like Virassamy, Minatchy, Govindin, and Ramasamy. Over time, however, these communities gradually lost the use of their ancestral languages, including Tamil and Telugu.

A few hundred from Calcutta were also introduced in the island. Kolkata was also a port of embarkation for indentured labourers from Malabar (coast) to Réunion Island. This word means the port is located in the present day in Kerala. And this was a nomenclature for all indentured labourers in Réunion.

A Huge Majority of Workers from Tamil Nadu

While the concept of *Girmitya* is well understood in many parts of the Indian diaspora, it remains relatively unfamiliar within the French-speaking diaspora. It is possible that an equivalent term once existed. In Tamil tradition, migrants were often referred to as "fathers" or "brothers", using terms like *kappa annin* or simply *kappa*, which conveyed respect and kinship. Meanwhile, Muslims from Nagore to Karikal became known in regions such as Sri Lanka, Malaysia, and Singapore by the term *Whahappa*.

Now if we go back to the year 1663, it is the year of the settlement of the first French people in Bourbon (knowing that the Arab merchant who particularly sailed the South of the Indian Ocean had knowledge of the island). The first Indians introduced in Bourbon were war prisoners sent in 1672 by Admiral Blanquet de La Haye, the viceroy of the East India Trading Company. They were deportees. The year 1687 deserves to be mentioned because on May 27 a Portuguese monk, Reverend Père Domingue de la Conception resold a 12-year-old slave, that he bought in India, to a Creole of the island, Gaspard Cautret. This young Indian is considered by all as the first imported slave.

It is only decades later that the governor Dumas (for Réunion island and Mauritius) went to Pondicherry in the name of the East Indian Trading Company to recruit labourers. Despite this approach, the adding of some slaves, craftsmen and other free workers reveals that the relationship with India remains relatively modest in this populating phase Les Etablissements français en Inde au XIXè siècle (1816–1914), Thèse pour le doctorat d'Etat d'Histoire, préparée sous la direction du.

During the early construction of the colony, skilled craftsmen from South India were brought in to contribute their expertise. They introduced traditional building techniques, including the use of *argamast*—an ancient method that served as an alternative to cement. Remnants of this technique can still be seen today in some of the island's older walls. Many Creole houses were constructed in the architectural style of Pondicherry, reflecting the influence of South Indian design. Officials and directors from the East India Trading Company, who had previously worked in Pondicherry, later settled in the colony and gradually contributed to the evolution of its local architectural identity.

In Barachois, a central square in the capital of Réunion Island, stands a prominent statue of famous Bertrand-François Mahé de La Bourdonnais, holding in one hand the treaty of occupation of Madras. Much could be said about his role and military campaigns in India. De La Bourdonnais also served as governor of the "sister islands"—Réunion and Mauritius (then known as Île de France). This chapter serves as a brief reminder that, during the eighteenth century, a contingent of Creole volunteers from Réunion was also sent to fight in India.

The history of indentured workers through sugar cane is widely known in the three French overseas departments: Martinique, Guadeloupe and Réunion. In the early eighteenth century, under the administration of the East India Trading Company, some affluent "Malabars"—part of the group known as *Libres de couleur* (Free People of Color)—began to establish recognized surnames or first names. However, the spelling of these names was often altered or misspelled, likely reflecting the phonetic conventions and orthographic norms of the French language at that time.

The island tried other crops like coffee, cloves, etc. There was also an attempt to produce silk. In 1841, seven years before the abolition of slavery, in a report published on the progress of the company of Mr. Perichon at Salazie: "The silks spun at Mr. Carré of Salazie, by a Malabar, were estimated in Lyon at the rate of 50.55 and 60 francs per kilo. For professional spinners were specifically from India: Lagime, Soubany, Hall, Scheck".

At the same period, Victor Robin from Saint-Pierre established at his place, Terre Rouge, a silkworm farm and two spinning tools to work according to the Indian process. In the nineteenth century, the first Indian immigrants were sent to Réunion Island. According to Captain Birsh, a police commissioner in Calcutta, migration to both Mauritius and Bourbon (now Réunion) was already underway by 1829.

Hugh Tinker, along with local historian Gérard Moutien Canabady, echoed an idea once considered by the grand planters: it was deemed "unwise" to place a cane knife—a symbol of power and autonomy—into the hands of individuals who had previously been subjugated, chained, and mistreated.

An Arrival Hard to Determine

For a long time, Réunion did not know the arrival date of the first batch of Indians. In 2022, after arduous research, the official date was discovered. On 3 June 1828, "La Turquoise" docked at *Le Barachois,* with 15 Telinga farmers on board, who had set sail from Yanaon on 16 March of the same year.[4] Since then, associations have commemorated this date in front of the monument dedicated to Indian workers in Saint-Denis. Although slavery had not yet been abolished at the time, Indian labourers were brought to the island as free individuals.

For the subsequent period of indentureship, the available information becomes more detailed and precise. At the end, a controlled immigration was set up before the abolition of slavery and was effective for several years (SINGARAVELOU 1994). The local authorities had time to conceive an international agreement, with England and France in 1860 for bringing the Indian workers on the island for work.[5]

The island's population figures reflect the significant transformations taking place during this period. Just months after the abolition of slavery, census data showed a rapid increase in the number of inhabitants.

While the large-scale arrival of Indians became especially noticeable after 1848, it is evident that the groundwork was laid much earlier. As early as 1820, Charles Desbassayns initiated a policy aimed at developing sugarcane cultivation, which began to yield its first results. At the same time, the work of engineer Wetzell brought about substantial technical advancements that further supported this agricultural expansion.

Who were the first indentured workers? This question has already been the subject of numerous scholarly contributions, making it unnecessary to revisit in detail here. Except the fact that some said Indians of high cast

[4] *Feuille Hebdomadaire de l'île Bourbon,* June 7, 1828, Archives departementales de La Réunion.

[5] International Convention of 1 July 1860 between England and France.

slipped into the convoys,[6] most of them came from the countryside's of South India, from the North or again from the popular districts of cities like Madras. Despite certain persistent oral traditions, the first indentured workers arrived on the island with very few personal belongings. Most had only a pair of trousers and a shirt, while some wore nothing more than a simple loincloth. Over time, a rich imagination has led some to believe that their ancestors brought along statues or treasured books, but there is little evidence to support such claims.

Some authors have described the journey itself. During the winter months, many passengers—encountering the sea for the first time—suffered greatly from the cold. Determined to deliver them safely and earn their full payment, ship captains often provided blankets to protect the migrants from the harsh conditions.

Yet the hardship did not end with the voyage. After ten years or more of labour, many workers returned to Madras or Pondicherry empty-handed. When possible, Indians migrated with their families, and laws eventually required ship captains—and later shipowners—not to separate them. Unfortunately, these regulations were not always followed. Families were often torn apart, with relatives ending up in distant parts of the island—for example, one member living in Saint-Denis in the north, and another in Saint-Louis in the south.

The Indian population in Réunion quickly surpassed that of Africans, largely because, from 1859 onward, England imposed strict restrictions on the recruitment of labourers from Africa.

The Contribution of Counters and Other Places of the South

	Pondicherry (soc. of mig.)	*Mahé*	*Cocanada*	*Yanaon*
1849–1860	46,129			1,300
1861		76		
1882			382	

(continued)

[6] The Darsanesing (Darsha) family claim that their grandfather was none other than Ranjit Singh's great-grandson, and that he fled from the British Râj, who had set out to eliminate all the sovereign's descendants. So, he came to Réunion Island. His research to back up his claims is ongoing.

(continued)

	Pondicherry (soc. of mig.)	*Mahé*	*Cocanada*	*Yanaon*
				1,300 (unknown date)

Two decades later, the volume of Indians recruits was formally estimated to have grown to around 6,000 per year. Even though the sub-continent cannot always fill out the "quota", it was hardly encouraged by the reports that were multiplying on the living conditions of those first expatriates. In fact, the planters were unwilling to get rid of slavery. The ill-treatment was regular and Indians employed under the contract attempted to resist but without success. They were mostly illiterates, and were not fluent in the local Creole language in a country that does not speak English.

Culture and Religion: The Preservation Levers of the Community

For a long time, many Indians in Réunion lived with the belief that it was forbidden to give their children Indian first names if they were born on the island. Wishing not to offend the sensitivities of the white population, and believing that a French name—particularly one reflecting the dominant Catholic religion—would help ensure their child's success, they often chose names accordingly. As a result, many sons of indentured workers were named Jean, and many daughters, Marie. Although this practice is increasingly challenged today, it still persists to some extent.

It is also worth noting that plantation owners, concerned with maintaining productivity, often allowed Indian labourers to observe aspects of their cultural practices during rest periods. The engravings of Roussin, a well-known painter of the era, depict various festive scenes from the early nineteenth century. In our own research, we uncovered a letter from the mayor of Saint-Benoît dated 1849—shortly after the abolition of slavery—that mentions a Tamil folk performance, known as *Nadagam* or Tamil theatre, illustrating the presence and endurance of cultural expression during that time.

The owners and the authorities realized that *Pongol* was one of the indispensable/*sine qua non* conditions of the coming of Indians on the island.[7] Another reason lies in the very meaning of the festival itself. *Pongal* is a harvest celebration, symbolically linked to agricultural prosperity—an idea closely aligned with the interests of plantation owners. Naturally, the landowners welcomed such traditions, as they reinforced the value of farming and productivity on the island. Even when the festival involved animal sacrifices at the start of the sugarcane harvest season, they showed no opposition, recognizing the cultural significance and its alignment with their economic goals.

The strong religious freedom still enjoyed today by the descendants of Indians in Réunion can be traced back to the earliest discussions surrounding the use of Indian labour. As early as 1827, labour contracts explicitly included provisions guaranteeing the possibility of this freedom.[8] However, the progress of these festivities is far from being identified and accepted, so easily by the public opinion.

> They are all followers of Vishnu-Brahma, some people are Muslim, but all also refractory to the catholic missionaries' affirmations as well as for those of the Protestants. Their religion, at least here, involves some noisy ceremonies on the occasion of the death of one of them, marriage or some other acts of their civil life, such as the piercing of ears or nobility.

Did the Indians Save the Colony?

Some observers have long questioned whether the Indians saved the economy of Réunion Island. While it is a serious and important question, it has often been addressed superficially, without relying on concrete evidence. *Le Mémorial de La Réunion*, for example, offers an implicit answer—suggesting, through a striking anachronism, that the colony might have experienced a better economic outcome had it not relied on Indian labour.

The authors wrote that "the immigration did not save the island but contributed to maintain techniques and mode of exploitation and especially a state of mind inherited from the slavery period. If there was no immigration, Réunion island would not have been economically lost, it

[7] A memorandum asks four days to be paid. Miot report, p 219—An essential condition for the dvt of the Indian immigration on the island.

[8] The contracts of commitment signed on the Coromandel Coast (Art. 8-Yanaon in 16.03.1827).

would have been better to use the arms it owned". This assertion does not resist for a long time to the reality of facts. We do not have to set up the indentured workers as savior. When we speak of indentured workers we generally think of Indians, but we do not ignore that other ethnic groups were among these labours. But it seems obvious that the contribution of the Indians was crucial in the economic development arising on the island. The settlers reported themselves that the presence of indentured workers was beneficial. D'Esménard,[9] protector of the immigrants raised that "*Indian Coolies saved the colony after 1848. Unfortunately, since then bad years happened. The immigration slowed down. It even came down to almost nothing. The renewal of workers became irregular. Subsequently we have to re-employed the old ones who took all the Creole flaws without even a ounce of their quality. It became almost impossible to have domestics and those whom they found were either drunkards or thieves*". Pierre de Montforand[10] strengthened this analysis by adding that, "*the Indian Coolie, if he was well-advised, well-surrounded, becomes an great worker (…) The Indians saved the colony after 1848. Indian indentured workers do not hesitate, in case of conflict with their employers to go on strike or to make complaints to the local authorities or to the British consulate*". Our author makes a nuance between immigration and Indians when he added immediately, "*fall-back position, solution of laziness, immigration, seen a century after, can hardly be considered as 'salvage' for Réunion*". "*The foreign workforce especially Indians to whom we resorted, brought prosperity for a moment, in particular between 1860 and 1870*".

The End of Indian Immigration

Indian immigration came under threat due to repeated warnings from the Viceroy of India, conveyed through his emissaries.[11] On November 11, 1882, Indian immigration to Réunion came to an abrupt halt.

[9] D'ESMENARD A., Les travailleurs étrangers à La Réunion in Album de l'île de La Réunion, Les pp.91 à 95, typographie de G. Lahuppe, Paris, 1883.

[10] MONTFORAND Pierre de, Société des Sciences et Arts, p 165, 1861.

[11] One of the PIO, Person of Indian origin, Egata-Patché Abady has taken legal action to have *"engagisme"* recognized as a crime against humanity, in the same way as slavery. His fight, which began in 2015, is taking place at both justice and research levels. Conferences and exhibitions have already been held on the subject.

However, the system of *engagisme* (indentureship) continued. Its implementation had required immense effort, to which many Indian workers contributed—some even losing their lives in the process.

Between April 27, 1848, and December 6, 1882, approximately 80,000 Indians—either British subjects or citizens of the French Empire—were brought to Réunion. This international agreement came to an end at the initiative of the British Raj, effectively putting a stop to further Indian immigration and placing the indenture system in a state of suspension. Nonetheless, indentureship lingered into the early twentieth century, sustained by the labour of ageing Indian workers already present on the island.

Britain's decision to terminate the agreement was driven by mounting complaints from Indian subjects about conditions abroad. Moreover, in an era of increasing maritime and commercial rivalry, Britain was reluctant to continue supplying labour to its major competitor, France.

Ties between India and Réunion gradually weakened. Most agricultural workers who had settled on the island chose to remain, especially since many had already started families. Fewer than a third of the labouring population opted to return to their ancestral homeland. Many of the former indentured workers requested parcels of land from their employers—even the most barren plots—seeking to cultivate and improve them. Sharecropping became a common practice.

Others established themselves as traders, artisans, or shopkeepers. By the turn of the twentieth century, former indentured labourers had made significant inroads into these sectors. A favourable law concerning the citizenship of foreign-born children also marked the beginning of the integration process for the descendants of indentured workers.

Over the decades, traditional lifestyles began to evolve. Until 1946—the year Réunion became a French department—it was still common to see sons and grandsons of immigrants wearing the *veshti* (a traditional loincloth), while women often wore saris in daily life. Today, such attire is mostly reserved for cultural or religious celebrations, but it remains a visible link to their heritage.

What became of the Indians at the end of their indenture contracts? Not all chose to remain on the island, and not all returned to their homeland either. Yet their presence endured, even grew, to the point where many eventually held prominent roles in Réunionese society—particularly in agriculture, where numerous individuals became small landowners.

In reality, after the cessation of immigration, a significant number chose to stay in the colony. Those who did laid the foundation for new family lineages, especially among those who had acquired citizenship, including emancipated former slaves and free workers who had completed their contracts. Over time, through inter-ethnic marriages, the community known as the *Malbars* came to represent a significant demographic. Many of those who settled were granted small plots of land, which they cultivated and made productive—playing a vital role in the island's agricultural development.

In the meantime, paradoxically many Indians had established a family in the island and bought a parcel of land. They had great interest in the country because in India they were abandoned by force of circumstance, families ties etc. Many believed it would have been difficult to adapt themselves to their native country. The planters would have been mistaken to oppose the settlement of Indian labourers, especially since they were not responsible for the end of immigration. In fact, many planters expended great effort to maintain the status of Indian workers as *coolies*—perpetually bound to labour. However, Great Britain opposed further development of the French colony, given the longstanding rivalry between the two nations. Ultimately, this strategic move served to protect British interests while limiting the growth of a competing French territory.

The island was well developed. There were more than a hundred factories on the island in the middle of the nineteenth century, and today two of them are still the jewel of the local economy.[12] In the end, the economic transformation of Réunion Island unfolded without major disruption. Indian labourers played a significant role, often through subcontracted work, alongside African, Madagascan, and Chinese workers. Together, they rose to meet the challenges of sugarcane cultivation—a sector that would come to define the island's modern identity as Europe's outpost in the Indian Ocean. Today, Réunion produces nearly 190,000 tons of sugar annually, most of which is exported for European consumption.

[12] *Bois-Rouge* and *Le Gol* are Réunion's last two sugar factories.

Indians Shaping the Social, Cultural, and Economic Landscape of Réunion

There is nothing easier than to make an inventory of the Indian presence in Réunion. It was true that the Indians were not the only one to build up the island but their contribution has been substantial. It is widely acknowledged that Indian labourers played a crucial role in both the development and preservation of Réunion's sugarcane industry. Today, Réunionese of Indian Origin (ROI) remain active in this sector, working in both the cane fields and the sugar factories. Building on the legacy of their ancestors, they have brought renewed energy and innovation to the industry. Like those before them, many have aspired—and succeeded—in owning parcels of land, with some acquiring several hectares.

Indentureship officially ended in 1882. In less than 25 years, old indentured workers conquered important parts of the land tax until they become the main owners in the leeward region. They took the lead of a great part of the sugar world in Réunion Island starting the end of nineteenth century. From 1895, Mourouguinpoullé afford to own more than 200 hectares from Million Des Marquets family in the East side of the island. "*Excepted Bois-Rouge, most of the lands belong to Malbar in the leeward region*" according to an observer, A. Marimoutou. The Indians are at every level of the sugar cane building. One cannot ignore the contribution of Angama Paniandy whenever there is mention about people like Joseph Mourouvin.[13] Both men were good friends and undertook numerous common affairs. One of their great achievements that they negotiated was the acquisition of the sugar refinery of the Ravine Glissante in the city of Sainte-Rose.[14]

Just a few years after the implementation of the immigration agreement, some Indians began expressing a clear desire to fully integrate into local society. While many remained on the margins, those who sought integration often found their own paths forward. As early as 1868, some had already established themselves as farmers and were employing

[13] Joseph Mourouvin can be considered the largest landowner, PIO, in Réunion. He owned more than 200 hectares of land, and even a forest, which he still owns today. His grandson is none other than politician Jean-Paul Virapoullé.

[14] Paniandy's son. A famous merchant in Réunion, "If there was no Paniandy, there would have been no Mourouvin", René Kichenin likes to precise. The grandsons or greatsons are in large numbers in agricultural unions. It is still true that the very delicate post of leader-cooker in the sugar refinery is still occupied by an Indian.

labourers. By 1876, Indian planters were emerging as notable figures in the agricultural sector. Later on, others ventured into urban services, particularly in drainage and sanitation, taking up the daily collection and maintenance of sanitary systems in the towns.

If smear campaigns were led against the Asians, some rare well-known people dare to venture into the trial of the indentured labours who integrate with difficulty into the Réunion society. Several Réunionese of Indian origin asserted themselves in the field of the public transportation at the beginning of the century.

Former indentured workers and permit holders demonstrated a strong aptitude for trade. Many among them sought to transition into commerce, hoping to improve their economic standing. Their ventures met with mixed success—while some managed to thrive, numerous others struggled. Many small shopkeepers, after just a few years in business, were unable to meet financial obligations and had to abandon their efforts. Several of these failed ventures ended up in court, often the result of inexperience or lack of clear direction.

Interestingly, some former *engagés* eventually rose to become landowners themselves, employing Indian labourers in turn—effectively reversing the roles they once held. One of the most symbolic examples of Indian economic presence can be found on Maréchal Leclerc Street (formerly Grand Chemin) in Saint-Denis. This major thoroughfare in the administrative capital was, by the early twentieth century, home to many Indian shopkeepers and traders. Though underexplored in historical research, various testimonies confirm that this street served as a commercial hub for Indian entrepreneurs. It is no coincidence that a Tamil temple was erected along this road—a testament to the deep roots of the community. Nearby lies the district of Saint-Jacques, known for housing a large population of descendants of indentured workers.

In another realm of influence, who in Réunion is unfamiliar with *carry* (curry), the staple dish of the island? Alongside *massalé*, several other culinary traditions were introduced from the Indian subcontinent. As for temples, they are so numerous across the island that attempting a full list would be exhaustive and incomplete.

Interestingly, the term *Kâla Pâni*—a significant concept in Hindu culture symbolizing the crossing of the "black waters" and loss of caste—does not appear in local Réunionese archives. Unlike in Mauritius or other Indian diaspora destinations, there seems to be no trace of this belief among the Coolies of Réunion, at least according to available oral traditions.

Beyond indentured labourers, the migration also included free passengers—craftsmen and artisans—especially towards the end of the *engagisme* period. These individuals further enriched the island's cultural and economic fabric.

Social and Cultural Contributions of the Indian Diaspora

In the end, we return to our guiding question: How has the Indian diaspora influenced the development of Réunion Island?

One compelling illustration lies in the legacy of Jacques de Châteauvieux, a descendant of early settlers who went on to lead the island's sugar industry. Today, he is recognized as one of the most prominent business figures in France—a true "son of sugar" who symbolizes the enduring economic impact of sugar cultivation, a sector shaped in no small part by Indian labour and expertise.

Wherever Indians have settled, they have marked their presence through the construction of religious and cultural landmarks. One remarkable example is the Tamil temple of Casernes in Saint-Pierre, which was officially designated a historical monument in 2007. Built around 1877, its façade still bears Tamil inscriptions, including the name of the goddess *Sakti Mariamman* and those of its benefactors—silent yet powerful testaments to the island's Indo-Tamil heritage. Similar fragments of Tamil script have been found scattered across the island.

The influence of Indian languages has also seeped into Réunion Creole. From Tamil, we find words like *rougail*, *tangol*, *ajoupa*, and *langouti*. From Hindi, words such as *kabay* (clothing), *dig-dig*, and *amaré* have entered everyday use, reflecting deep-rooted linguistic and cultural exchanges (FIRMIN 1986).

The imprint of India even touched some of Réunion's most notable literary figures. Fanny, the mother of poet Auguste Lacaussade, owned two slaves—one of them from Bengal. Leconte de Lisle, another renowned poet, expressed his fascination with India in his poem *Le Manchy*, where he references the *Telinga* workers (Telugus from Andhra Pradesh). And during a brief visit to the colony in 1840, Charles Baudelaire was so captivated by an Indian woman that he immortalized her in his poem *Une Malabaraise* from *Les Fleurs du Mal*.

We may temporarily conclude this overview of Indian influence in Réunion with a symbolic gesture: in 2016, the word *curry* (or *kari*) was officially added to the Larousse Dictionary, affirming the dish's place not only on the island's tables but also in its cultural and linguistic identity.

We are also interested in developing diasporic literature. Mauritius and Réunion have contributed a lot in French literature, and the contributions have been mostly from Indian settlers.[15] Also, we have plans for translating the French literature of the Sugar islands into Tamil, and Tamil literature into French.

There is no question of engaging in an inventory of products that relate to India, but simply to highlight some lines that appear and highlight the contributions of Indians in the Islands cultural and social history. The island has adapted itself to rice, the national dish, which is undoubtedly, the contributions of the Indian Indentured labourers. The soaring of the price of rice in the world market did not deter the islands from importing rice, it has been acculturated, though locally the islanders preferred bread over rice but it remains the special dish when inviting a guest, this certainly exemplifies the importance of rice on the island way of life.[16] We also have, apart from Rice, spices like curry (a form of curry), and different varieties of vegetables that are the contribution of Indians in the islands. Every locality has their vegetables markets on different days, which can be traced back from the times of the indentured labours.[17]

Regions such as Guadeloupe, Martinique and Réunion island, where most people belong to the Indian diaspora are from Tamil-speaking areas. They use Tamil for prayers, perform the same rituals that they have been practising for centuries, perform the kavadi, and walk on fire, which are all rituals attributed to the Goddess Mariamman, and know 50 to 60 tamil words. These people want to learn tamil as it is a part of their heritage. The problem is to teach Tamil to Tamils who left their homeland nearly 150 years ago. In Réunion, Hinduism is the second largest religion. It is also a significant sign of diversity.

[15] Joseph Varondin, the poet Anne-Marie Gaudin de la Grande corresponded with Rabindranath Tagore. But they never met. Professor Gilles Sagodirana Pillai has found one of her precious letters.

[16] Around the 1970s, the people of Réunion were still consuming Mangalore rice.

[17] Note that products as yams, maize, or other starchy foods were consumed by slaves or engaged but disappeared.

On the cultural and religious aspect, it is not necessary to look very far. It is enough to mention that we have over 20,000 people attending *Diwalî* in each local celebration during the months of October–November every year.[18] Over the past 50 years, this celebration has become the flagship or the torch bearer in popularizing the Indian cultural heritage on the island. But today we have different groups that oppose, especially from the government circle. The *Pongol* (originated from Tamil Nadu), Sankranti, is still marked in temples or in some families. The ***kavady*** (Tamil festival to Subramanian, son of Shiva), is the subject of a large mobilization, as well as walking on fire... In the latter event, the faithful want to remember the pleasant call for budding actors, the scenes of the Mahabharata. However, tamil theater *(Nadagam or Nalgon)*, ***karmon,*** (tamil folklore) are threatened with extinction.[19]

Finally, the Bollywood phenomenon encountered a fertile ground in Réunion Island. Many young people have identified with the explosion of Indian fashion, seeing a homecoming through many Indian channels like Zee TV and many more. Besides the elegant Sharukh Khan, it is popularly remembered thanks to his concert on the island and by promising to make movies on the island as the islanders were emotionally attached to the Bollywood icons.

During a stay in Mauritius, Krishna Menon, former Indian diplomat from the Ministry of External Affairs, recalled that he met a former chairman of Réunion Island and told him this observation "*Michel Debre is not Réunionese, but Réunion Island belongs to him*".[20]

Conclusion

Does India Still Influence the Development of Réunion Island? The relationship between India and Réunion today is quite different from that of the past. While the historical ties remain strong, the dynamics of influence have shifted. One can recall that just two years ago, the

[18] M. Virapoullé Jean-Paul, municipality of St Andre, former mayor and deputy, got a Sâman Award, PBD (2006), because of the popularity of the Dîpavalli (Dîwali) festival, which attracts around 20,000 people each year.

[19] Since the end of 2022, at the request of Région Réunion, the Bal tamoul (Nadagam or Nalgon) has been included by the Ministry of Culture in the national inventory of intangible cultural heritage.

[20] This elegant quotation tends to show the colonial link that still existed at that date.

regional council—Région Réunion—initiated discussions with Indian Oil to explore potential cooperation. Unfortunately, those negotiations did not come to fruition.

Nevertheless, symbolic and cultural connections continue to be honoured. In 2014, organizations such as GOPIO, ODI, IDC, alongside local associations and the Consul General of India, came together to commemorate the pivotal contributions of Indian ancestors. A monument was erected at Le Barachois as a lasting tribute to the indentured labourers who helped shape the island. The ceremony was graced by Mahesh Sharma, then Indian Minister of Culture and Tourism, who became the first Indian cabinet minister to officially visit Réunion Island—an important milestone in Indo-Réunionese relations.

From the first day when they put feet on Réunion Island ground, Indians begin to conceive strategies to settle there. Moreover, not all the Indians come through indentureship. Let us take the example of Gnanavisayen Tambymestry who received in 1857, an "*invitation to provide boarding for India, in its favour*" from the Medicine naval school of Toulon (France). We can quote several examples[21]:

For the vast majority of Indian descendants at the beginning of the twentieth century, the priorities had shifted. Education, or rather the obligation to attend school, was increasingly seen as a gateway to stable employment—a means to secure long-term livelihood. The pursuit of social advancement became closely tied to academic success.

During this period, a strong sense of community solidarity emerged within certain Malabar families. One notable example is Antoine Narassiguin, who benefited from the unwavering support of his elder, Gabriel Sinanapoullé. Thanks to Sinanapoullé's guidance, Narassiguin obtained his basic school certificate in 1914, and a few years later, with continued encouragement, he went on to earn his higher certificate.

Jean-Baptiste Varondin thanks to his education with the Jesuit Brothers of Sainte-Marie, is the first Réunionese of Indian Origin holder of the school Certificate. Moreover, at the brothers' school, he obtained a

[21] In 1906, Gnanadicom, a state employee came from the French India (Pondicherry), was working in Saint-Pierre, in the judiciary. Also, let us mention the case of Asservathen Ephra, merchant, who was born in 1862 in Madras (English India), living in Saint-Pierre, on "decision of the president of the Republic, on February 6th, 1893 was authorized to settle in France and to enjoy civil rights there, as long as he will keep on living here ".

accounting degree, very much sought after during this period, which brought him to work with personalities such as Emile Hugot.

In the same lineage, we can quote Antoine Soubou, also a school director. In 1935, he was 14 years old when he received his school certificate. He attended then the teachers' training college. He obtained an upper certificate in gymnastics. He was successively director of the complementary course in Saint-Paul, labour activist, a local councillor still in the same city. Today, a middle school is named after him in Saint-Paul.

In reality, the real caesura took place in 1946 with the departmentalization. As Prosper Eve underlines, with the advent of the status, Réunion especially had to favour the assimilation. So, the diverse demonstrations of the foreign representation were relieved in the background. It is not useless to remind that the change announced in 1946 was not applied immediately. It was necessary to wait several years before noticing the first profits of the law sacralizing the former colony to a French department.

Freed from their contracts, many former indentured workers succeeded in acquiring portions of land that were far from negligible. For instance, Joseph Mourouvin came to own nearly 6,000 hectares, a considerable figure when compared to the 30,000 hectares initially held by Gabriel Le Coat de Kerveguen, one of the island's largest landowners.

While thousands of Indians were deployed to European fronts during the wars, their relatives in Réunion were not passive observers. Antoine Narassiguin, for example, served in Noyon during the First World War, where he narrowly escaped death on the battlefield. After suffering injuries and enduring a difficult return voyage, he eventually made it back to Réunion. His struggles didn't end there: during the Second World War, in 1941, he fell victim to the oppressive regime of Marshal Pétain, and was dismissed from his teaching position.

Historian Singaravelou recalls a disturbing remark made by Henri Sidambarom in Guadeloupe:

> "In a century, there will be no sons of Hindus of pure blood left in this country".

Could it be that sentiments like this contributed to a deep discouragement among indentured workers and their descendants, deterring them from passing on their cultural and religious heritage?

Indian Identity

During slavery, the owners showed no big regard on the question of food given to the slaves. But under the indentureship, the new relationships led them to supply some rice to Indian workers. The colony not producing this cereal anymore they had to engage negotiation with Vietnam, Madagascar, and India to import it.

The question of assimilation or integration was also in the food debate. Having made sure that their religion would not be forbidden in the colony, the Indians saw their culinary practices taken into account. Some indentured employers served corn and manioc to their workers instead of rice had to answer to the violation of the rule sometimes.

The rice file highlighted the increase of the Indo-musulman importance in Réunion. At the beginning of the century, in spite of their more recent settlement than their indentured fellows, Indian storekeepers, native mainly from Gujarat, were motivated to integrate Réunion island society. We also see that this integration was made with some reluctances, even rejection from a certain part of the population.

Despite these episodes of compulsory adaptation to the crisis bound to the war, the rice remained a safe element of the local cuisine. The Indian indentured workers group, and then the Réunionese population had to import this foodstuff. It is here a particular integration formed because the committed did not need to impose their lifestyle. At the end, their food habits and then their culinary habits (carry) were adopted by all.

Vestige

During renovation work on the Cathedral of Saint-Denis in the year 2000, a remarkable discovery was made. Hidden on the back of a wooden ledge fragment, workers found letters inscribed in Tamil. This unexpected find confirmed that one or more Indian craftsmen had contributed to the cathedral's woodwork during its construction in 1858. The DRAC (Regional Directorate of Cultural Affairs) documented the inscriptions, which are believed to be the initials or markings of the Indian artisans who left their subtle yet lasting imprint on this historical monument.

The generalization of the practice of the Tamil ceremony shows that indentures and their descendants benefited freedom in leisure activities. From Saint-Benoît to Saint-Paul or Saint-Pierre, the descendants held high the flame of this folklore which their ancestors immortalized.

The double religious practice goes back to the dawns of indentureship history in Réunion. The cult of *La Salette* in 1881 was so important that an observer estimated the indentures was about 10,000 to participate. They offered gifts to *Notre-Dame-de-la Salette* (candles, oil). Besides, they also thought they would be granted healing by offering food to the pilgrims. This appropriation of La Salette's days was little appreciated by certain religious persons in charge who perceived a certain "depreciation" of the cult which found its origin in the miracle that happened to the children Maximin and Mélanie.

Before even the adoption of the great agreements, settlers seized the opportunity for them to authorize the popular cults of the Indian workers. An inverse decision would have probably limited the Indian immigration policy. While contracts were set over the decades, temples—despite the reluctances of the Church—build up in diverse Réunion municipalities. It was the first stone sealed in the project of Indian settlement here. With the blooming of the numerous religious buildings, it became unthinkable that all the Indians left massively the colony.

Finally, here is an illustration with this anonymous testimonial.

> The Indians took out their large tank and came in procession to have the traditional bath in the River Saint-Denis. The crossing of the city, by the street Dauphine, was made without the beat the tam-tam under the supervision of two police officers. Except for the shouts "Coinda". There was in the escort and among those who pulled the rope of the tank, many more Creoles than the real Malbar. It is not need to add that the walk in the city had a big success. All the terraces, all the bars were full of curious people.

Finally, in order to deepen the understanding of the Indian contribution to French national history, it would be appropriate for the term *"engagisme"* to be officially recognized in the French dictionary. This notion is gradually gaining acceptance.

References

GERBEAU Hubert, « Des minorités mal connues, esclaves indiens et malais aux Mascareignes au XIXe siècle », Communication présentée au Colloque de Sénanque, mai 1978. p 161.

LACPATIA Firmin, « Quelques aspects de l'insertion des Indiens à La Réunion au XIXe siècle ». p. 319. Actes de la Conférence Internationale organisée par l'AHIOI, France-Inde, (Saint-Denis de La Réunion, 21–28 juillet 1986).

SINGARAVELOU, « La diaspora indienne dans la Caraïbe : essai de bilan », in Présence de l'Inde dans le monde, Ensemble d'études par l'auteur, p.83. Gerec/Puc, l'Harmattan, 364 pages, 1994.

WEBER Jacques, *Les Etablissements français en Inde au XIXè siècle (1816–1914),* Thèse pour le doctorat d'Etat d'Histoire, préparée sous la direction du Prof. J-L Miège, soutenue en mars 1987, publiée par la Librairie de l'Inde, en 1988, 5 tomes.

Complicating the Narrative of Hindu American Activism: The Case of the Hindu American Foundation

Jeffery D. Long

The Hindutva Question: Serious Analytical Tool or Political Bogeyman?

The basic thesis of this paper is that viewing the Hindu American Foundation, or HAF, through the lens of its location in its American social and political context is far more revealing and adequate to its highly complex subject matter than a study that views this organization solely as an extension of Indian political movements and organizations.

Why is it necessary to argue for such a seemingly self-evident thesis? What other conceptual lens would one use to examine an organization established to advocate for the interests of a religious minority community in the United States than one attentive to the American political context and to prevailing attitudes towards religious minorities in that country?

This need arises from the fact that the Hindu American Foundation, in the pursuit of its advocacy aims, has become the target of allegations

J. D. Long (✉)
Religion, Philosophy, and Asian Studies, Elizabethtown College, Elizabethtown, PA, USA
e-mail: longjd@etown.edu

A. Dubey and S. Rajwar (eds.), *Diaspora in Development*,
https://doi.org/10.1007/978-981-96-6747-5_5

that it is nothing more than a front organization for the Hindu nationalist movement in India, and that its sole purpose is to advance a Hindu nationalist agenda in America.[1]

Why is being labeled a Hindu nationalist or Hindutva organization a problem? If the label fits, what is the issue? The problem here has two dimensions: one substantive and one rhetorical. The substantive issue–that the agenda of HAF is simply not identical to that of the Hindutva movement–will be the main focus of this paper. The rhetorical issue has to do with the intent behind allegations of Hindutva; for it is not the case that those who utilize these labels always see them as mere neutral descriptors of an ideology.

Hindu nationalism, defined as an ideology aimed at formally proclaiming India a Hindu, as opposed to a secular, state, or *Hindu Rashtra*, is widely associated in the minds of many with intolerance and violence–rhetorical and actual–against religious minority communities in India. Whether this is a fair characterization of the entirety of the Hindu nationalist movement, or only of certain extreme elements at its fringes, is debatable, and is also beyond the focus of this paper. In terms of intellectual respectability, though, the point is that to be successfully labeled a Hindu nationalist is, at least in many quarters, the proverbial kiss of death–a point on which I shall elaborate below.

To be sure, there are significant areas in which the interests of at least some parts of the Hindu nationalist movement and the Hindu American Foundation have coincided. This is understandable, given that, whatever their differences, both the Hindu nationalist movement and HAF see themselves as advocating for the interests of Hinduism and the Hindu community. It is also the case, as critics of the Hindu American Foundation have alleged, that there are persons in HAF with ties to Hindu nationalist organizations.[2] The aim of this study is not to contest these facts, but rather to look beyond them to the wider context in which the HAF has emerged.

[1] Such allegations can frequently be found on social media, such as, for example, in the comments on the Hindu American Foundation's Facebook page, or in conversations on the Religion in South Asia listserv (prior to a decision by the moderators of this listserv in December 2015 to probhibit commentary on news items shared by list members).

[2] Kurien, p. 159. It is interesting to note, though, that Kurien only names one member of HAF, Mihir Meghani, with close ties to Hindu nationalism in India. Meghani, a founding member of HAF, is certainly a prominent voice within the organization. It is not clear, though that he is a typical HAF member in having his particular affiliations.

To return to the question of the point of the allegation that a given organization or some of its members are affiliated with Hindu nationalism, in this paper, all references to Hindu nationalism and to Hindu nationalist organizations are descriptive references to an ideology and to the organizations and individuals that openly and proudly adhere to that ideology. But in the very heated discourse surrounding the Hindu American Foundation, to give an organization or an individual the label *Hindu nationalist* or *Hindutva* is a move that seems to be intended to discredit, and thereby silence, the voices of those with whom those who apply such labels disagree.

Such usage is thus akin to labeling a person or a group as *fundamentalist* (a label that is also applied by some to Hindu nationalists, and thus, by extension, to the Hindu American Foundation, by those who engage in such labeling). Once a person or a group has been so labeled, everything with which that person or group is associated becomes suspect. Efforts to "expose" the Hindu nationalist affiliations of a person or a group thus become akin to the proverbial witch hunt. If one can discredit a person or group through guilt by association, one is spared the effort of actually having to engage the perspective of that person or group. Discourse of this kind is, in the judgment of this author, not befitting of serious scholarship. Again, this is distinct from the question of whether any particular person or organization does, as a matter of fact, adhere to a Hindu nationalist ideology (the neutral, descriptive sense that I use in this paper). Are there real overlaps between the Hindu American Foundation and the Hindu nationalist movement, in terms of both ideology and membership? There are indeed.[3] Should such overlaps exhaust our understanding of what the Hindu American Foundation is and does? The claim of this paper is that answering this question in the affirmative leads, at best, to incomplete and inadequate understanding of the Hindu American Foundation and its aims and ideology, and is, at worst, misleading and intellectually dishonest.

[3] The number of HAF members, though, with a formal affiliation to a Hindu nationalist organization is quite small. I have heard it estimated to be as low as one or two persons– far less than the number that might be mentally conjured by the phrase "some members of HAF have Hindutva affiliations." See previous footnote.

Full Disclosure

Before proceeding further, in the interests of candor and full disclosure, I should mention, as already noted in my abstract, that I served as a consultant for the Hindu American Foundation on a number of issues in the early 2000s. In the process, I also developed good, collegial friendships with many members of HAF. These relationships have enabled the kind of access that has permitted me to pursue the research on which this paper is based.

I should also note, though, that my consultation work has been entirely voluntary and free. I have never been in the employ of the Hindu American Foundation.

Finally, and this last point is extremely important in a context in which politically charged allegations are being made, I do not always agree with the HAF on every single topic on which it makes a formal pronouncement; nor do the members of the HAF always agree among themselves. The HAF is not a monolith. It is a dynamic organization that is comprised of persons with diverse backgrounds, perspectives, and life experiences.

Overview of the Argument

The argument of this paper has three main parts. The first involves clarifying the meaning of the term *Hindu American*. Both Hindu nationalists and their critics often fall into the error of conflating Hindu and Indian identity, of conceptualizing *Hinduness* as an ethnic category, rather than a religious one. Such a conflation is an error, at least in the American context–a context informed by the largely Protestant Christian cultural matrix of the United States–in which Hinduism is a religion, and to be religious is not to be part of an ethnic group, but to be a voluntary member of a like-minded group of people, based on certain shared values, beliefs, and practices. Even as individual members of the Hindu American Foundation might question the suitability of the term *religion*, as opposed to *dharma*, for what they see as an authentically Hindu self-understanding, nevertheless, the Hindu American Foundation operates within the American discourse in which a Hindu is a self-identified adherent of Hinduism, regardless of what that person's ethnic identity or national origin may happen to be. This fact immediately problematizes claims that the HAF is an extension of the Hindutva movement in India, for the understanding of what it is to be Hindu in its American context is radically different from the concept of a Hindu nationhood with which the Hindutva movement generally operates.

The second part of the argument draws upon the words and experiences of HAF members. Rather than merely casting labels about, such as *Hindutva* or *fundamentalist*, what picture do we get when we are attentive to the words of the people who make up the Hindu American Foundation? What are their self-understandings regarding the aims and ideology of the organization they support, and in some cases, of which they are founding members? What becomes strongly evident as one delves into the interviews I conducted with HAF members is that we are hearing voices from a minority community. To be an advocate for Hindu interests in India is something that can certainly raise suspicions, as reflecting a majoritarian ideology, akin to that of conservative Christians in America. It is important to be mindful, though, that in the United States, it is Hindus who are in the minority. Language indicative of Hindu advocacy that might be inflammatory in India is, in an American context, arguably quite appropriate.

Finally, the third part of the argument draws attention to HAF initiatives that have placed this organization at odds with the Hindu nationalist movement, thus demonstrating definitively that a simplistic identification of the Hindu American Foundation with Hindu nationalism is precisely that: simplistic, and inadequate to the complex reality in which this organization operates. Our aim here, again, as the title of this paper indicates is to complicate the narrative of Hindu American activism, not to oversimplify it, much less to demonize it with politically charged labels.

Before developing this argument, though, a brief overview of the history of HAF is in order.

What is the Hindu American Foundation? History and Context

After the lifting in 1965 of the Asian Exclusion Act of 1924, which had banned most immigration from India to the US, growing numbers of Hindus moved to the United States for both employment and education. Hindus from India and from other parts of the world began, in the seventies, to build temples and to become a regular part of the social fabric of America. The tech boom of the nineties increased this trend.

Coinciding with this influx of "ethnic" Hindus, of Hindus who had been born into and who grew up practicing Hindu traditions, there also emerged a growing fascination with Hinduism and Hindu thought and practice among the majority population. Practices such as meditation, yoga, nonviolence, and vegetarianism, as well as concepts such as reincarnation and religious pluralism, held a great appeal for many Americans, especially those involved in the counterculture of the sixties, but for many since that time as well.[4]

By the middle of the nineteen nineties, sufficient number of Americans identified themselves as Hindu for American Hindus to see themselves as a distinct group within the larger American societies, with distinctive interests and concerns. Prominent among these concerns was the representation of Hindu thought and practice in academic writing emerging from American colleges and universities, as well as in middle school and high school textbooks. After the terror attacks of September 11, 2001, which was followed by a wave of Islamophobia that has yet to abate fully, there was also a concern in the minds of many Hindu Americans to differentiate Hindus and Hinduism, in the perception of the average American, from Islam.[5]

It was in this context that the Hindu American Foundation was first established on September 3, 2003, in Fremont California. The vision of this organization, as expressed on its website, is "To sustain a leading institution of Hindu American advocacy for the promotion of dignity, mutual respect, and pluralism." This combination is significant. "Hindu advocacy," in an Indian context, is a majoritarian agenda. But the emphasis here is on "mutual respect" and "pluralism"–on Hindus as one group among many in the United States.

The identity of the Hindu American foundation is further elaborated in its mission statement:

> The Hindu American Foundation (HAF) is an advocacy organization for the Hindu American community. The Foundation educates the public

[4] Prominent studies of this phenomenon include Philip Goldberg's *American Veda: From Emerson and the Beatles to Yoga and Meditation–How Indian Spirituality Changed the West* (Harmony Books, 2010), Lola Williamson's *Transcendent in America: Hindu-Inspired Meditation Movements as New Religion* (NYU Press, 2009), and Paul Oliver's *Hinduism and the 1960s: The Rise of a Counter-Culture* (Bloomsbury, 2014).

[5] I recall in this connection a conversation that was recounted to me by one HAF member who was meeting members of the US Congress in order to educate them about Hinduism and Hindu issues. The first question that one Congressman asked her was "What is the difference between Sunni Hindus and Shia Hindus?".

> about Hinduism, speaks out about issues affecting Hindus worldwide, and builds bridges with institutions and individuals whose work aligns with HAF's objectives. HAF focuses on human and civil rights, public policy, media, academia, and interfaith relations. Through its advocacy efforts, HAF seeks to cultivate leaders and empower future generations of Hindu Americans.[6]

This mission statement clearly differentiates Hindu identity from that of an ethnic or national group, thus distinguishing the Hindu American Foundation's understanding of Hindu identity from that of the Hindu nationalist movement, which identifies Hindu-ness with Indian-ness:

> The Hindu American Foundation is not affiliated with any religious or political organizations or entities. HAF seeks to serve Hindu Americans across all sampradayas (Hindu religious traditions) regardless of race, color, national origin, citizenship, caste, gender, sexual orientation, age and/or disability.[7]

The activism of HAF has been directed at a variety of issues that the organization takes to be of central concern to Hindus. These have included the aforementioned issues of representation of Hinduism in school textbooks, but also the human rights abuses that have been suffered by Hindus globally, domestic American issues that are related to the separation of church and state (like attempts to erect displays of the Ten Commandments on public property, or anti-Hindu comments by persons running for office), abuses that are related to caste in India, and the representation of yoga as an exercise routine with no relation to Hinduism or Hindu spirituality (controversially called the "Take Back Yoga" campaign).

The issues on which the HAF has taken a stand cut across the political spectrum, in both Indian and American contexts. In its advocacy for revisions of school textbooks that present Hinduism in a negative light (while maintaining a fairly positive image in their representations of other religions), as well as in its "Take Back Yoga" campaign, the interests of HAF and the Hindu nationalist movement in India would seem to align quite closely. At the same time, in its extensive critique of casteism, HAF found

[6] http://www.hafsite.org/about-us/who-we-are, accessed 18/9/16.

[7] Ibid.

itself running afoul of many of the same Hindus who had cheered its activism in the textbook cases. Its advocacy for separation of church and state in the United States places it firmly on the side of political progressives who consistently challenge attempts to transform the United States into a "Christian nation." The HAF is neither the RSS nor the ACLU, yet its agenda and interests overlap with those of both of these organizations. The reality of HAF is clearly a complex one, not reducible to a single factor or ideological orientation.

Who Is a Hindu American?

Who are the Hindu Americans that the Hindu American Foundation claims to represent? It is important to clarify this question, for its answer in an American context is quite different from that which might be offered in India, not only by adherents of Hindu nationalism, but also by their critics. In India, one finds less reference to *religion* and to *religious* affiliation than one does to *community* and *communal* affiliation. This is not a merely semantic difference. The concept of *community*, as deployed in India, has a set of connotations not unlike the concepts of *race* or *ethnicity* in the United States. How does one join a community? This is typically a matter of birth, in which the individual person does not exercise choice. One does not choose one's community, but simply *is* part of it.

Consequently, if one has favorable or unfavorable feelings or views regarding one community or another, this has direct social and political implications, often of a highly unpleasant nature. To express criticism of a community is to express prejudice. The law, in a democratic polity, must ensure that persons do not experience discrimination on the basis of the community to which they belong, and the rights of minority communities, being particularly vulnerable, are in need of special protection.

This overlaps with, but is also in important ways quite different from, the concept of *religion*. As developed in a largely Protestant Christian context, *religion* implies a set of beliefs and a way of life to which one adheres because one chooses to do so. On this understanding, to be part of a religion is not so much a matter of birth (though people do tend, overwhelmingly, to adhere to the religion of their parents). A religion is less like an ethnicity and more like a club. One chooses a religion and joins a religious organization because one agrees with its teachings and adheres to its values and practices.

Again, these two concepts–community and religion–overlap in numerous ways in practice. Religious groups tend to act like communities. We even label them accordingly as *religious communities*. And statistically, people do generally adhere to the religion of the family in which they were born.

At least as an ideal type, though, to hold a positive view of a specific religion–to agree with and to assent to its teachings and worldview–and to hold a negative view of a different religion–to disagree with its teachings and worldview–is not to participate in a prejudice of the kind with which a person may be charged who likes or dislikes members of a particular community. To adhere to one religion and not another is simply a logical and, I would argue, inevitable consequence of having a worldview at all. We all believe things: about what is most important in life, about the ultimate aim of our existence (and whether our existence even has an ultimate aim), about what happens to the conscious entity after the death of the physical body, about whether there is a deity or deities, and what we ought to do in response this fact, and so on. And we all tend to affiliate and to identify with persons who share our views on these topics, whether our worldview is of a religious or a secular nature. Disagreement with a worldview only becomes a prejudice if it begins to impinge upon how one feels and behaves towards persons who hold that view. Disagreement is inevitable, but it can be expressed respectfully or disrespectfully. Thus it is not Islamophobic or Hinduphobic, for example, to disagree with the core concepts of Islam or Hinduism. Prejudice is against persons, not ideas.

The Hindu American Foundation operates in a cultural milieu where the concept of religion is distinct–certainly in a legal sense–from the Indian concept of *community*, which, I would argue, functions in ways closer to that of the concepts of *race* or *ethnicity* in the United States. This is an important point, for some Indians and Indian Americans have criticized the Hindu American Foundation for being an exclusionary (and of course Hindu nationalist) organization precisely for not defining itself as an *Indian* American Foundation. The thinking behind this criticism, quite understandably, is that Indians are a community within the United States–an ethnic group comparable to the communities of India–and this is certainly the case. And because Hindus are seen as a subset of this community, and because not all Indians (or Indian Americans) are Hindu, advocating for Hindu interests is seen as exclusionary and as unnecessarily divisive.

If, however, in the United States, to be a Hindu is a matter not of ethnic, but of religious affiliation, then this criticism is misguided; for if being a Hindu is a matter of adhering to a Hindu worldview and spiritual practice, then anyone can be a Hindu. And in America at least, this is, in fact, the case. A recent Pew research poll that studied the ethnic make-up of American religious communities revealed that, while the vast majority of Hindus in America are, in fact, of Asian heritage–I think we may presume that the overwhelming majority of these are either from India or of Indian descent, though many are no doubt from other South Asian countries and other parts of the world as well–not all are. Among the roughly 2.23 million Hindus in America, the Pew study found that 91% are of Asian heritage. 2% identified themselves as being of "mixed" heritage. (We will probably be safe in presuming that most of these have one parent who is of South Asian descent.) 4%, however, identified themselves as white, 2% as African American, and 1% as Hispanic. This means there are approximately 89,200 white American Hindus, 44,600 African American Hindus, and 22,300 Hispanic American Hindus. These numbers are not insignificant. They suggest that non-Indian or non-South Asian Hindus, while yet a small minority in the Hindu American community, are a definite presence within it.[8]

The woeful inadequacy of a simple identification of Hindu-ness with Indian-ness is affirmed, and is in fact presumed, in the above-cited line from the Hindu American Foundation's mission statement: *HAF seeks to serve Hindu Americans across all sampradayas (Hindu religious traditions) regardless of race, color, national origin, citizenship, caste, gender, sexual orientation, age and/or disability.* This line would be incoherent if Hindus were all of a single race, color, and national origin. HAF presumes a Hindu American community that exhibits characteristic American diversity in terms of ethnicity and national origin. This is very different from the fundamental assumption of the Hindu nationalist movement that Hindu-ness is coextensive with Indian-ness.

The non-Indian or non-South Asian presence in the Hindu American community is also not merely a phenomenon at the periphery of this community, and thus something that might be easily dismissed as insignificant. As Lola Williamson and Ann Gleig have shown, some of these

[8] http://www.pewresearch.org/fact-tank/2015/07/27/the-most-and-least-racially-diverse-U-S-religious-groups/ft_15-07-23_religiondiversityindex-1/, accessed 18/9/16.

non-Indian or non-South Asian American Hindus have taken up leadership roles within the Hindu American community.[9] There are of course the gurus that Williamson and Gleig describe, such as Satguru Sivaya Subramuniyaswami (born as Robert Hansen)–the founder of a publication, *Hinduism Today*, that is an important voice within the emergent global Hindu community–as well as numerous non-Indian swamis and pravrajikas of Hindu or Hindu-based organizations such as the Ramakrishna Order. Many of these non-Indian Hindus are from the majority white or European American ethnic group; but this is not exclusively the case. There is, from the African American community, Bhakti Tirtha Swami (born as John E. Favors), who was a prominent leader in ISKCON (the International Society for Krishna Consciousness), and from the Hispanic American community, Sri Dharma Pravartaka Acharya (born as Frank Morales).

Finally, non-Indian leadership in the Hindu American community is not confined to swamis and acharyas, but also householders, and is become increasingly normalized. I can note my own participation as a member of the Board of Trustees at my local Hindu temple near Harrisburg, Pennsylvania: the Hindu American Religious Institute, or HARI. My nomination and election to this role (which was not something I had solicited or for which I had sought) occurred without opposition and passed with unanimous affirmation from the community.

Given this context, the Hindu American Foundation sees itself as representing the interests not of an exclusionary subset of the Indian American community, as critics may allege, but rather of the highly inclusive and ethnically diverse emergent Hindu American community, bound not by ethnicity, but by religious belief and practice.

In confirmation of this self-understanding, in answer to the interview question, "Why a Hindu American Foundation and not an Indian American Foundation?" the answers that I received included the following: "The question could be asked—why a Federation of Indian American Christian Organizations of America, or an Indian American Muslim Council, or a Sikh Coalition or a Jain Foundation? Religion is primary and supreme in the American consciousness and in the political, cultural and social arenas, and being a majority Christian country, minority Hindus need to become agents for informing the rest of society about the Hindu

[9] Ann Gleig and Lola Williamson, eds., *Homegrown Gurus: From Hinduism in America to American Hinduism* (Albany: State University of New York Press, 2013).

faiths, philosophies, traditions, and cultures so that we can respond both positively in constructing and sustaining a diverse society but also step in to protect and defend the rights of Hindus in their new homeland." Another respondent said, "I think many Hindus born and raised in the United States care more about their identity formation in the United States than in India. Moreover, I think Hindus as a diverse religious minority in the American social fabric need to be identified as such. It also is a disservice to the hundreds of thousands of Indian-Americans who are *not* Hindu." And in the words of one respondent, who is of Indian descent but born in the United States, "I don't think Indian American defines me the way that Hindu American does. Indian and American are at times identities that are at odds with each other. I simply can't wear jeans and a sari at the same time. On the other hand, I've never once felt like my Hindu identity competed with my American ethics. The Indian identity is something that my parents owned and passed down, while the Hindu identity is something I've developed myself and come to own."

Of course, the vast majority (91%) of Hindus in America are also ethnically South Asian (mostly from India). There are thus many issues on which most Hindu Americans can and do unite with the greater Indian American community. In many instances, for example, in which both Hindus and Sikhs face harassment or physical violence in the US, this is due, ironically, to Islamophobia on the part of their attackers, who typically do not differentiate among these groups, and may not even be aware of their separate existence.

This is because an American bigot who is bent on harassing or attacking someone different from themselves will not usually be sufficiently knowledgeable to differentiate between a person with brown skin who is Muslim and one who is Hindu or Sikh. Being a bigot, such a person will also be highly unlikely to care one way or the other. This sad phenomenon of indiscriminate Islamophobia-motivated attacks on all Indians is clearly a strong reason for Indian Americans to work as a united group. An attack on one is an attack on all, regardless of religious affiliation. Nothing, however, precludes a Hindu of Indian descent from supporting the work of the HAF and at the same time supporting and participating in the work of organizations of South Asian or Indian Americans as well. None of these organizations demands exclusive allegiance. My personal observation is that it is not at all uncommon for Hindu Americans to belong to many organizations with overlapping, but distinct, aims and purposes. And HAF has condemned all religiously motivated violence, not only violence against Hindus. In many ways, HAF models itself on Jewish advocacy in the United States, focusing primarily, but not exclusively, on the rights of one particular community.

In Their Own Words: HAF Members Speak for Themselves

How do HAF members respond to questions about their organization, its ideology, and its aims? I should note that in compiling this research, I was able to speak with HAF members from a range of backgrounds. The persons whose words are presented below included persons who were born and grew up in India and migrated to the United States as adults, persons who were born in India but who migrated with their families in early childhood, persons whose parents are from India but who were born and have lived their lives in the United States, and one white American Hindu (not myself). The ages of the persons interviewed ranged from their early twenties to seventies. Importantly, the interview group included a roughly equal number of men and women.

In response to the question, "What drew you to work for (or to help establish) the Hindu American Foundation?" the following answers are of particular interest:

"I was born in the US, and though my parents have worked hard to help me understand and develop my Indian and Indian-American identity, I've always felt a lot more comfortable with the Hindu piece of my identity, because it actually describes who I am, what I practice, and what I believe. Since it's the strongest piece of my identity, and one that's made me a better person, I was drawn to HAF and the idea of helping other Hindus develop pride in their identity. I've been very frustrated of late with peers who take a very passive approach to developing that pride and scoff when others try to help others develop theirs, and through HAF I've found a group of like-minded people who work hard to give the community said pride." This respondent affirms the distinction between an Indian identity and a Hindu identity. Both are parts of who she is, but she is most comfortable with, as she says, the Hindu piece of her identity. This answer is also significant in affirming the fact that we all have multiple identities. None of us is simply reducible to ethnicity or to a religion; and in American context, a religious identity, as a thing chosen and reaffirmed by the individual, as opposed to an accident of birth, can be more central than one's ethnicity in one's self-understanding.

Another respondent came to the Hindu American Foundation through her pursuit of grassroots level interfaith work. Her response highlights the character of the Hindu community as a minority community in the American context: "In early 2005, I was in the midst of a controversy over the National Day of Prayer in my hometown in Michigan. The city was going to exclude or segregate faith communities for its annual National Day of Prayer ceremonies. I was on the front page of major newspapers, as the Hindu woman who spoke out against exclusion from this event…The numerous calls we had during the following years as I became an interfaith activist and speaker for the Hindu community made me understand the importance of HAF's advocacy work, and also realize how much I needed like-minded Hindus who understood the challenges. Interfaith dialogue as it stands right now is dominated by Abrahamic traditions, and more about finding common ground between "People of the Book." People speaking about Hinduism would say things that are not necessarily based on fact or knowledge of Hinduism as a lived religion, based on normative Hindu practices, and some—within interfaith circles—also misrepresented, demeaned, or diminished the tradition. I also had children who went to middle school and high school during these years, and saw them go through the same kind of frustration I had gone through during my childhood. My father, now a retired English professor and writer, had advised me during the same time that Hinduism was being seen through a Western lens, and taught me how to explain our practice of Sanatana Dharma. I found that the leaders of HAF…all used similar language and had a similar understanding of the challenges we faced with exclusion and misrepresentation of Hinduism, and knew how to speak up with the nuance required."

Contrary to depictions of the Hindu American Foundation as a mere extension of the conservative Hindutva movement in India, one respondent answered this question in language characteristic of the progressive American left: "I joined because I wanted to be able to bring together my spirituality with my goals of social justice." Younger members in particular, who have grown up in the US, often have little interest in Indian politics.

In response to the question, "Why do you feel that Hindus, specifically, are in need of advocacy of this kind?" one interviewee said, "Lots of people ask questions, which temple leadership doesn't always have answers to. What is the Hindu position on abortion, care of the environment, animal rights, etc.? HAF and its leaders have written about various issues and provided both official policy level briefs as well as the diversity of Hindu perspectives on issues including abortion, gay marriage, gun rights, animal rights, caste, religious freedom, and proselytization/conversion. Temples and other Hindu institutions can reference these documents. The viewpoints are also helpful to Hindus who may not be affiliated with any organization or institution but want an articulate way of explaining the issues. My daughter in 9th grade was told by her teacher that the Abrahamic faiths were the original monotheists. Long story short, she reached out to HAF leadership, the Academic Advisory Council and others in the community to explain the misunderstanding—that Hinduism is both polytheistic and monotheistic, and that the western way of looking at or describing, and the history around the study of Hinduism comes from a place that often diminishes the tradition. Having role models at HAF as well as the resources and connections available through HAF will help her and other young Hindus be strong in their pursuit of dharma." This interviewee also spoke of the tendency for Indian political issues to be brought into the American arena of discourse, saying, "There are [also] negative forces at play—[some] in the Muslim community who speak loudly about the rights of minority religious groups in India, certain Sikh leaders who want to lay the violence against Sikhs in India in 1984 at the door of all diasporic Hindus, and Christians who didn't like my questions regarding predatory proselytizing. A strong and strategic Hindu voice that is reasoned and respectful, expecting mutual understanding and fair treatment, is needed to advocate against these negative forces and promote pluralism."

Another respondent to this question spoke of the need for fair representation and inclusion in American society, "Every group of people in the United States deserve a voice, and HAF is ours. That said, we are in more dire need than other faith groups, because we are incredibly misunderstood. The average American's piecemeal picture of a Hindu is exotic and hard to relate to. We have much work to do before people understand the very basics of our faith, and that we aren't so very foreign after all."

Because parallels are sometimes drawn between the Hindu American and Jewish American experiences, I asked the question, "Do you see parallels between the work of HAF and the situation of Hindus in the United States and the work of organizations such as the Anti-Defamation

League and the situation of the Jewish community in the US?" The answers to this question included the following, "I believe there are certainly parallels between any group that's been misunderstood and discriminated against on the basis of beliefs. The Jewish community has done a stupendous job of taking control of their own narrative, and the Hindu community must learn from them and apply their successes."

A quite provocative response to this question was, "HAF's work is critical to reduce Hinduphobia—similar to what the Jewish and other minority religious groups face. Hindu self-loathing similar to Jewish self-loathing is also an unspoken reality, something I see even when I am at the temple. In fact, people who follow what I see as Hindu practices–both those in and out of the temple–rarely like to call themselves Hindu, based on my experiences."

Regarding the phenomenon that this respondent characterizes as "self-loathing," which manifests as persons who follow Hindu practices and adhere to a Hindu worldview not wishing to identify themselves as Hindu, I have observed it as well, and written about it elsewhere.[10] In some cases, this phenomenon does indeed appear to be a result of a kind of self-loathing, or shame, resulting from the experiences of marginalization and ridicule to which many Hindus are subjected. In other cases, though, it arises from impulses that are internal and integral to specific forms of Hindu thought towards identifying not with a limited concept from the realm of *nāma-rūpa*, but with the *ātman*, or universal self. Out of such philosophies arise the phenomenon in which many people who might otherwise be regarded as Hindu instead make statements like, "My religion is nonviolence" or "My religion is truth." A similar philosophical orientation characterizes the distinct but related phenomenon of the "Spiritual But Not Religious" movement in the United States: a sense of any identification with a finite identity–ethnic, national, or religious–as an impediment on the spiritual path.

Finally, though, there is also an aversion among many to the kind of politicization to which Hindu identity has been subjected in India, such that identification as Hindu becomes conflated with adherence to the Hindutva or Hindu nationalist movement. For many, it is better to set aside the term *Hindu* than to be identified with such an ideology.

[10] "Hindu-To Be or Not To Be: Three Possible Reasons for Aversion to the Term 'Hindu' among Western Practitioners" (unpublished paper presented at the Dharma Academy of North America Annual Meeting in Chicago, October 30, 2008).

How do members of the Hindu American Foundation respond to this concern? In response to the question, "What is your response to the criticism that the Hindu American Foundation is an extension of the Hindu nationalist movement into the American political scene, and that HAF is a Hindutva front organization?" a number of interesting responses were provided. One respondent said, "It is interesting that Indian Christian, Muslim, Jain, and Sikh organizations are not asked if they are fronts for proselytizing, monopolistic, and other enterprises, including political ones. Those who took the initial lead to support and speak for Hindus happened to include members of the RSS and the VHP and given the fact that those two organizations had already been demonized in India by a whole host of Indian political and religious groups, it was easy to label Hindu activists here as Hindutva activists." Another said, "It's laughable, so it's not worth a response. I think there has been a concerted effort to attach a stigma to the word Hindu." According to another, "The first time I was accused of being a Hindu nationalist or Hindutva, I had to ask my husband, who was raised in Hyderabad, what it meant. I could sense it was derogatory, based on how it was used. I had been engaged in building community and seeking social justice for many years prior to joining HAF, wanting people to stop "othering." I had nothing to do with any kind of politics. It was all about my local and regional community, and helping weave the various aspects of who I am and what I come from, into my own identity. I am not Hindutva and neither is HAF a Hindutva front organization, if by that one means [it] seeks to make all Indians Hindu, or make India a 100% Hindu nation...Several of us at HAF have been involved in various Hindu organizations—we work with people in these organizations due to an interest in the mission of that organization. None of the work that we do as part of these other organizations is related to or seeks to influence Indian politics. Just as HAF is non-partisan and critical of any elected official or public figure in the United States who says something in contradiction to our vision to promote pluralism and advocate for the Hindu community, we use the same rules when it is relevant to speak about issues in India or elsewhere in the world." Finally, another said, "The Hindu American Foundation has no interest in the creation of a Hindu nation, which is what I understand a 'nationalist' movement to be, and nothing that we have said or done comes anywhere near implying we would ever be in support of such a thing."

The Hindu American Foundation at Odds with Hindu Nationalism

Finally, the Hindu American Foundation has pursued a number of initiatives, such as its study on caste and its statement in support of marriage equality, which have placed it at odds with at least some portions of the Hindu nationalist movement.[11] Regarding initiatives such as these, HAF members have said, "Hindu American Foundation clearly has a strong track record of taking independent positions—leftists and radicals throw the label Hindutva at HAF, and right-wing Hindus call HAF knee-jerk liberals! Yet the reality is that HAF is neither. The team at HAF thinks through positions that have nothing to do with the usual left-right paradigm present in India or the US, and therefore HAF rejects any labels."[12]

The HAF report on caste–*Hinduism: Not Cast in Caste*–released in 2011, evoked a considerable backlash from many Hindus who saw it as an attack on Hinduism and who argued that HAF was a front group for anti-Hindu forces in the US–quite an ironic charge when one also takes into account allegations from the other end of the political spectrum that HAF is a front for Hindu nationalism, due to its involvement in the school textbook issue in California. The brief on homosexuality, issued more recently, provoked much less backlash, although the issue of gay rights had emerged as a major dividing line in the Indian context. This more muted reaction was attributed by one HAF member whom I interviewed to the fact that Hindu acceptance of homosexuality has a stronger foundation in Hindu scriptures than the arguments advanced in the previous report on caste. Another noted humorously that those who might have otherwise been outraged by the report on homosexuality had in case already given up on HAF after the caste report.

[11] http://www.hafsite.org/media/pr/hinduism-not-cast-caste-full-report and http://www.hafsite.org/media/pr/haf-policy-brief-hindusim-and-homosexuality, accessed 18/9/16.

[12] http://www.hafsite.org/sites/default/files/Coalition_Against_Genocide_A_Nexus_of_HinduphobiaUnveiled.pdf, accessed 18/9/16.

Conclusion

The Hindu American Foundation exists within an American cultural and political milieu in which religious communities routinely advocate for acceptance and inclusion in American society, and in which minority religious communities, in particular, routinely argue for and affirm the values of tolerance and pluralism on which the United States was founded, but which it has embodied only imperfectly at each phase of its history: a prime example of what Jürgen Habermas calls the "unfinished project of modernity."

While it is true that, as a Hindu advocacy organization, HAF will naturally have overlaps with the Hindu nationalist movement, to the degree that both see themselves as advocating for the cause of Hindus and Hinduism, to reduce the HAF to an extension of Hindu nationalism into the American scene without attending to the varied ways in which this organization expresses the ongoing quest of Hindu Americans to define themselves–ourselves–in the wider multicultural reality of which we are all part is to do a disservice to this quest, and to the aspiration for greater inclusion, and ultimately, social justice, which informs it. This does not mean the Hindu American Foundation is immune to criticism, or that it cannot be called to task for specific stances that it takes or that its individual members take if representing themselves as speaking for it.

In the judgment of this author, HAF's willingness to advocate for the elimination of caste prejudice, even in the face of enormous criticism for doing so, as well as against homophobia, also with some resistance, illustrate that this organization is not simply a front for Hindu nationalism, but an internally diverse organization seeking to represent a similarly diverse community dealing with an array of complex issues that do not simply fall along the familiar political polarity of "left" and "right."

Work Cited

Hindu American Foundation and Mohammad Siddiqi, "Coalition Against Genocide: A Nexus of Hinduphobia Unveiled" (2013) http://www.hafsite.org/sites/default/files/Coalition_Against_Genocide_A_Nexus_of_HinduphobiaUnveiled.pdf, accessed 18/9/16.

Hindu American Foundation (2016) "HAF Policy Brief: Hinduism and Homosexuality" http://www.hafsite.org/media/pr/haf-policy-brief-hindusim-and-homosexuality, accessed 18/9/16.

Hindu American Foundation (2010) “Hinduism: Not Cast in Caste” http://www.hafsite.org/media/pr/hinduism-not-cast-caste-full-report, accessed 18/9/16.

Hindu American Foundation (2018) “Who We Are” http://www.hafsite.org/about-us/who-we-are, accessed 18/9/16.

Pew Research Center. “How Racially Diverse Are US Religious Groups?” http://www.pewresearch.org/fact-tank/2015/07/27/the-most-and-least-racially-diverse-u-s-religious-groups/ft_15-07-23_religiondiversityindex-1/ accessed 18/9/16.

Diaspora and Development: The Indian Community's Contribution to the Gulf's Economy and Society

Aditya Anshu

Introduction

The Indian diaspora in the GCC (Gulf Cooperation Council) member states, Saudi Arabia, the United Arab Emirates, Qatar, Oman, Kuwait, and Bahrain, is one of the largest and most diverse migrant worker populations in the world. Millions of Indians reside and work in the Gulf, and affect the economic, social as well as the cultural landscape of the region. The Indian population in the Gulf is multi-generational and comprises individuals from different states of India. Thus, representing a multifaceted nature of migrant life in the Gulf. They are associated with nearly all aspects of the economy as they are engaged in various professions ranging from construction to healthcare, finance to information technology etc. These migrants play a crucial role in infrastructure development, service delivery, and the human capital powering the economy of the region. Additionally, the contributions of the Indian diaspora are not only limited to monetary contributions and manual labour, but also influence the social and cultural scene in the Gulf nations. Indian culture, religion,

A. Anshu (✉)
Abu Dhabi University, Abu Dhabi, UAE
e-mail: aditya.anshu@adu.ac.ae; adideomishra@gmail.com

A. Dubey and S. Rajwar (eds.), *Diaspora in Development*,
https://doi.org/10.1007/978-981-96-6747-5_6

and traditions have significantly impacted the way of life in the Gulf. Indian cuisine, music, festivals, and social bonds have left an indelible mark on the region. However, the relationship between the Indian diaspora and the Gulf is not unidirectional and is also complex. Members of the Indian diaspora are also faced with challenges in the region, particularly with regard to issues of labour rights, living conditions, and the absence of permanence of stay or citizenship. The present chapter aims to highlight the multi-dimensional contributions of the Indian diaspora in the Gulf countries and examines the ways in which the Indian diaspora has impacted the development of the region. It also explores the challenges faced by the members of the Indian diaspora communities including visa limitations, labour exploitation, and challenges of assimilation. The present chapter focuses on the interaction between the agency of Indian migrants and the structural forces which condition the flow of migrations and diaspora life. Additionally, it aims to offer a comprehensive understanding of the place of the Indian diaspora in the Gulf and what it signifies for host countries and for India. In creating the context for the analysis, the chapter employs two theoretical paradigms: Transnationalism Theory and Migration Systems Theory. Transnationalism theory examines the mechanisms through which migrants connect their two worlds—India and the Gulf through the medium of economic exchanges, social relations, cultural interactions and significantly influence both these places. Migration Systems Theory offers an explanation of the economic, historical, and structural forces that facilitate and sustain migration flows. It describes the institutional and systemic pressures that play a role in establishing the pattern of migration, and creates an interdependence between the host and home countries. Overall, this present chapter provides insights into the considerable impact of the Indian diaspora on the Gulf. It explores the economic and cultural intricacies of such influence on migration, identity and development. Through an exploration of agency and structure within its framework, the chapter sheds light on the Indian diaspora's contribution to regional development and its implications for India's place in the world more broadly. By implementing these theoretical paradigms, the chapter will demystify the two-pronged effect of migration, gaining insight into the challenges and opportunities that are engendered within the diasporic setting.

The Historical Diaspora in the Gulf: A Historical Context

Trade and cultural contacts between India and the Gulf nations have been existent for centuries. Ever since the days of Indus valley civilization, India had trade relations with West Asia. The major routes for this trade included the Persian Gulf Route and the sea route extending along the Persian and Arabian coasts through Aden and into the Red Sea (Jain 2005). Even though Indian traders had been present in the Gulf since the Bronze Age, the first historical reference to an Indian community in the Gulf can be found in a book from 916 AD by the Arab historian Abu Zayd Hasan. In his account he had mentioned an Indian community of over 100 merchants in Siraf, Iran. The oldest existing Indian community in the Gulf is in Muscat and which is similar to the Hindu settlement mentioned by Abu Zyad Hasan. This community dates back to the fifteenth century. The second oldest community is at Manama, the first reference for which can be found in Dutch sources in 1667. However, it is very much likely that this community is older than that. The third oldest community is the Lawati community of Matrah and Muscat. Its origin can be traced back to the Ya'ariba dynasty (1625–1743), as proposed by J.E. Peterson. The individuals that made these communities came primarily from Sindh, Kutch and the Kathiawar peninsula (Onley 2014). Many of the Indians in the past were elite and wealthy members of the society and had a profound cultural and political influence on the region. Up until 1947, India and Indians were highly regarded in the region which gradually declined during the 'oil boom' of the 1970s and the consequent surge in the demand for labour. The Asian workers including the Indians were favoured because of their flexible work hours and persistent hard work. However, the migration from India to the Gulf declined in the 1990s. This happened as a result of restrictive immigration policies and localization policies imposed to control the non-Arab populations in the region (Nayudu 2022). The 1990–1991 Gulf War triggered a mass exodus of Indians from Kuwait, but following the country's liberation in 1991, many returned, making them the largest expatriate community (Singh and Arimbra 2019). In the early 2000–2008 period there was a resurgence in the demand for skilled and unskilled labour in the Gulf due to increase in the oil prices. Additionally, the Government of India too facilitated international migration through its policies. According to the UNDESA Migration Report 2017, 60% (16.9 million) of the 28 million foreign labourers in Gulf nations are South Asian, out of which Indians account for 31.5% of the total migrants. Nearly 70 per cent of the Indian migrants work as labourers or technicians in the construction sector, as domestic servants and drivers. The trend of labour migration from India to the Gulf countries is shifting from states like Kerala, Telangana, and Tamil Nadu to states like Uttar Pradesh and Bihar.

Economic Contributions of the Indian Diaspora: A Structural Perspective

Migration influences the economy of both home as well as host countries. The structural perspective in the context of migration of Indians to the Gulf countries explores the role of systemic forces such as historical migration trends, labour markets, recruitment networks, and government policies etc. influence the migration flows. Through the lens of migration systems theory the present chapter explores the sectoral and economic contributions of the Indian diaspora in the Gulf countries.

Understanding Migration Through Migration Systems Theory

Migration systems theory was proposed by Mabogunje (1970). It was later extended by Kritz et al. (1992). According to migration systems theory (MST), international migration as a process does not occur in isolation. Instead, it is influenced by interactions between different nation-states and a variety of other factors including socio-cultural dynamics, economic policies, and geopolitical conditions. Since these factors are external and not solely influenced by the individual choices, migration flows may be explained and predicted by understanding these factors. According to Fawcett (1989), three kinds of links between nations influence migration: relational, regulatory, and tangible. Relational factors refer to historical links, family ties, and cultural connections. Regulatory factors refer to laws and policies that regulate migration. These regulations determine who can migrate, under what conditions and for how long. For instance, 'Saudization' also called 'Saudi Nationalization Scheme' or 'Nitaqat' was introduced in 1985. It mandates the private businesses to hire a certain percentage of Saudi citizens. This may gradually reduce reliance on foreign labour. Another example is that of the 'incipient' status of the Indian diaspora in the Gulf (Jain 2005). Incipient migrants refer to a relatively sizable group of foreign workers in industrial or oil-producing economies who are ethnically distinct from the host population and who are 'allowed to remain in their host country only to work' but are not entitled to become citizens (Weiner 1982). The decision of how many migrants can immigrate depends primarily on the political factors and long-term power needs (Weiner 1982). Tangible factors refer to physical and economic connections between the nation-states that influence migration. Explanatory hierarchy and exposure hierarchy serve as ways to explain how these different factors at multiple levels affect migration. According to explanatory hierarchy, migration is not only determined by the characteristics of the individual nations but also by the shared similarities between the home and the host country. These shared characteristics facilitate the migration. Even though the primary language in the Gulf is Arabic, English serves as a common language. Many Indians are proficient in English. Moreover, due to the migration history, there is a high concentration of Malayali workers in the Gulf. This linguistic familiarity with English and Malayalam languages eases communication and facilitates migration of Indians to the Gulf. The presence of a well-established Indian diaspora in the Gulf also makes it easier for more Indians to migrate. Additionally, because of the trade history of India with the Middle East, including the Gulf Indian rupee was widely used in the Gulf for trade. During colonial rule, the British also used Indian rupee for trade in the Gulf as well as other countries such as Africa and parts of Asia. In the twentieth century many of the Arab sheikhdoms also used rupee as a legal tender. However, as the countries gained independence, they began to introduce their own currencies. By 1966, Dubai and Qatar had introduced their own currencies. By 1969, most Gulf states had replaced Indian rupee with their local currencies. In 2022, Sultanate (Oman) embraced tech money innovation 'Rupay' (Bhandare 2020). Some Gulf nations are exploring the possibility of integrating technology similar to Unified Payment Interface. The Gulf nations and India are also economically interdependent. While the Gulf nations supply India with a huge quantity of oil, India supplies the Gulf nations with food, textiles, and manpower. The oil boom resulted in rapid development of the region creating a huge demand for foreign labour which was fulfilled to a great extent by the skilled and semi-skilled Indian labourers. In addition to explanatory hierarchy, exposure hierarchy explains that the countries differ in their level of engagement in the migration flows. Countries adopt different roles in the migration process. Some countries assume a long-standing role as a country of origin, transit or destination country. Thus, making them core 'migration hubs' some countries may play only a minor role in the migration process or may have recently become a part of the global migration pattern. Countries that mostly serve as migration destinations are called 'core countries'. Countries that serve as both origin and destination are called 'semi-peripheral countries'. Countries that mostly serve as origin countries are called peripheral countries. In the context of the migration between India and the Gulf, the Gulf nations serve as core nations for Indians and India serves as a peripheral nation by providing a labour force to the Gulf nations. Gulf countries on the other hand provide employment and remittances back to India.

Role of Remittances

Remittances refer to the money sent by the migrant workers back to their home countries. These financial transfers play a crucial role in the economy of both the home and host countries of the migrants. In 2014, over $35 billion as remittances came to India from the Gulf region alone, making up more than 50% of total remittances (Anjum 2017). While remittances bring a lot of benefits, they also create certain challenges. Not only do remittances provide immediate financial relief to the receiving homes in India but also contribute to long-term development. Children from remittance-receiving families tend to seek higher education, which brings in a better skilled labour force in the future. Healthcare services are enhanced because families spend money on higher-quality nutrition and medical care, resulting in a healthier and more productive workforce. Nevertheless, this dependence on remittances can also lead to economic vulnerabilities. If relatives rely too much on foreign funds rather than contributing to local employment, then the workforce can dwindle, leading to reduction in productivity and innovations. Foreign exchange can also contribute to inflation. Thus, causing the living expenses to rise. As a result, fundamental commodities can get out of reach for those who lack remittance subsidies. Occasionally, countries may also experience currency appreciation which is commonly known as the 'Dutch Disease'. A study by Faheem et al. (2022) reveals that migrant remittances can lead to an appreciation of the real effective exchange rate (REER). The study provides evidence for the existence of Dutch Disease risk in India since a rising currency can render Indian exports more costly and less competitive in the global market. It involves appreciation of the local currency due to the surge in remittances. Thereby reducing the export competitiveness. For the host countries, migrant workers can bring economic stability by filling up the labour shortages. Moreover, migrants also contribute economically to the host nations in the form of living expenses and taxes paid. Also, a significant amount of capital outflow can reduce the amount of money available for local investment and consumption. However, since the Gulf economies are primarily oil-driven, remittance outflows do not pose a major financial crisis but can affect liquidity in certain sectors. Additionally, in some Gulf countries the number of expatriates is higher than the locals. This has created a dual economy, one for the high-income earning locals and another for the low-income expatriates. Even though expatriates dominate the labour force, most of them are engaged in low-skilled labour (Cohen 2008). In order to reduce reliance on foreign workers, Gulf nations have introduced nationalization policies like Saudi Arabia's 'Saudization' and the UAE's 'Emiratization'. These policies are made to make job opportunities for the local people better by welcoming companies to hire more Gulf nationals compared to expatriates. However, since many locals prefer government jobs or higher-paying positions, migrant workers remain essential in labour-intensive industries.

Social and Cultural Contributions of the Indian Diaspora: A Transnational Theory

Transnational migration refers to the process of migration across international borders in which the individuals build and maintain varied connections with their country of origin while they try to settle in the host country (Fouron and Glick-Schille 2001). Indian diaspora in the Gulf exemplifies transnationalism. While the Indians try to assimilate and settle in the Gulf countries, transnationalism shapes their identity, cultural contributions and social networks (Biswas 2021). Methodological nationalism focuses on the experiences of the migrants while they try to adapt to the culture and society of the host nation. However, this often ignores the dual point of social, political and cultural reference and connections that these migrants maintain in their home as well as host countries (Upegui-Hernandez 2014). Many of the Indian migrants in the Gulf may be regarded as transmigrants. They lead their lives between national borders, economically engaging in the Gulf while at the same time socially, economically, and politically engaging in India. For instance, they may work in the education sector in Abu Dhabi but remit money back home to maintain their families or participate in the local elections. The concept of transnational social fields is also relevant to examine the social and cultural contributions of the Indian diaspora. It refers to the social spaces formed through people's social and political interactions between their places of origin and their home country. Members of the Indian diaspora share a common cultural background, maintain ties to India, and often create cultural organizations and community groups in the Gulf to preserve their traditions and social connections. Transnationalism can be exemplified in the 'Gulf boom' in Kerala which is characterized by mass migration to the Gulf countries from the state of Kerala, especially during the period of 1972 to 1983. This has significantly impacted Kerala's economy by injecting remittances, improving infrastructure and enhancing the standard of living. However, this also creates the problem of 'brain drain'. The Indian diaspora also plays a pivotal role in India's cultural diplomacy. Through business, cuisine, festivals, and language, the Indian diaspora creates cross-cultural interactions that reinforce India's diplomatic relations with the Gulf countries. The Indian diaspora has also formed a series of associations and organizations in their host countries to preserve Indian culture and languages. Indian expatriates have established renowned cultural, religious and professional organizations in the Gulf. For instance, there are six registered Indian Associations in Abu Dhabi/Al Ain. These include: India Social and Cultural Center (Abu Dhabi), Abu Dhabi Malayalee Samajam, Kerala Social Center, Indian Islamic Center, Indian Social Center (Al Ain), and Indian Ladies Association (Embassy of India, UAE). These associations extend support to the welfare initiatives and conduct cultural/welfare activities regularly including sports activities. These efforts have resulted in improved cultural ties and mutual relations between India and the Gulf countries.

It is also important to note that the leaders in the Gulf not only aim to use their wealth for the development of their country but also keep their political systems unchanged. They share wealth with their own citizens by providing social services and government jobs, but they do not share political power. Although some observers were concerned that migration would overwhelm the Gulf countries and threaten their regimes, Indian migrants do not oppose the existing regimes for the fear of losing their economic safety. As a result of the above approach to migration, Indian migrants often face restrictions and are forced to live in their own social enclaves. Because of their incipient status, Indian migrants cannot own land, become citizens, join trade unions, participate in trade unions and have limited social benefits. As a result, a social order divide has been created in the Gulf, between the productivity-oriented migrant population and consumer-oriented native population. Consequently, the region is witnessing economic growth without the local population incorporating the values associated with modern economic growth (Weiner 1982).

Labour Rights Issues and Exploitation Within the Gulf

The Indian diaspora in Gulf countries, especially the migrant workers, routinely experience horrendous labour rights violations and discrimination. Many workers are involved with construction, domestic work, and hospitality, where they endure miserable working conditions, long hours, and low pay. Kafala, an employment system practised in Saudi Arabia, UAE, and Qatar, ties the workers to the employer without giving them the power of changing jobs or even leaving the country without the employer's consent (Jones 2016). This system increases the risk of exploitation, with workers frequently complaining of non-payment of wages, substitution of contracts, and passport confiscation. The migrant workers also suffer from scant living conditions, being housed in overcrowded accommodation that offers limited access to basic facilities. Reports of physical and psychological abuses, especially on the part of domestic workers, underline the inadequacy of effective legal protection. Although the reforms occurred in some Gulf countries recently, the enforcement mechanism is weak and hardly covers what little rights the workers have, exposing them further to vulnerability (Auwal 2010). The absence of strong trade unions worsens the situation, putting the workers at a disadvantage in negotiating for their rights collectively. The challenges must be addressed through a strong regime of legal protection, international cooperation, and diplomatic pressure from the Indian government to ensure the well-being of its own citizens abroad.

Visa and Work Restrictions: Impact on Social Mobility and Worker's Rights

The visa regulations and work regimes severely restrict and hinder the social mobility of Indian workers in Gulf nations. Many migrant workers come on fixed-term contracts, limiting their opportunities for better wages. The kafala system empowers the employer to the extent that their control over the legal status of the worker makes it almost impossible for the worker to change jobs or contest exploitative conditions. Hence, most of the time, the workers remain trapped in low-paying, physically demanding jobs with very little scope for advancement. Family visa restrictions mean that very often, workers may spend years without being able to bring their families to be with them, which affects their mental health and well-being (Gardner 2011). The diaspora is even further isolated by the lack of permanent residency or citizenship in most Gulf nations, which prevents long-term social and economic integration. Policies to restrict the mobility of workers also restrict access to education and healthcare, which now requires financial backing from employers because most low-income workers cannot afford it. Visa reforms to facilitate greater mobility for occupation, pathways to long-term residency for skilled workers, and enhanced law enforcement against exploitative practices are urgently needed to change conditions for the better.

Political Engagement and Advocacy by the Indian Diaspora in Gulf Countries

Despite legal and political restrictions, the Indian diaspora in Gulf countries has increasingly engaged in advocacy efforts to improve their rights and living conditions. The laws in Gulf nations tend to restrict migrant workers from having unionized protests or engaging in any political discourse. Nevertheless, informal advocacy organizations, as well as social media have served as alternative platforms for expressing concerns against this injustice. The Indian government through its embassies and diplomatic missions, has also played a part in advocacy for workers' rights. Initiatives such as the Pravasi Bharatiya Divas and e-Migrate system are developed for the welfare paradigm of overseas Indians (Auwal 2010). Also, some powerful Indian business communities in the Gulf have a good rapport among the local authorities and act as mediators between the migrant workforces and the local authorities. Despite such handsome efforts, there still remains a wide gap when it comes to formal channels for worker representation. Strengthening engagement through diplomatic channels and international cooperation might assist in pushing for more comprehensive labour protections in the region. The political engagement and advocacy by the Indian diaspora in the Gulf countries significantly enhance India-Gulf relations by fostering economic interdependence through substantial remittances, improving labour agreements to protect worker rights, and promoting cultural diplomacy that strengthens India's soft power. The diaspora's active involvement in cultural and social activities creates goodwill and bridges cultural gaps, while their role in facilitating business networks and educational exchanges deepens strategic partnerships. Additionally, the diaspora serves as a vital link for political and diplomatic engagements, helping to resolve bilateral issues and enhance cooperation. This multifaceted involvement not only bolsters India's geopolitical influence in the region but also contributes to regional stability and mutual development.

Addressing the Need for Policy Reforms Related to Diaspora Welfare

Policy reforms are imperative to improving the conditions of the Indian diaspora in the Gulf countries, particularly the low-income migrant workers. The regulation of recruitment agencies is one such sector that needs immediate policy changes. Most of these agencies partake in exploitative practices like charging huge fees and providing false job contracts. Additionally, enhancing the pre-departure training and awareness of workers regarding workers' rights and legal frameworks in host countries would benefit them. Bilateral agreements with Gulf countries should be expanded to bring about effective compliance with progressive labour laws regarding wages and grievance mechanisms. Furthermore, the Indian government must create more robust legal aid and welfare schemes such as the Indian Community Welfare Fund (ICWF), which provides money and legal aid to distraught workers abroad (Abraham 2012). More welfare-enhancing social security measures with pension schemes and reintegration programs for returning workers would go a long way in improving the long-term welfare of the diaspora. By prioritizing these policy reforms, India can ensure that its citizens working in the Gulf receive better protection and recognition for their contributions to both economies (Naha 2024).

Challenges and Future Outlook in vis-à-vis Indian Diaspora in Gulf

The Indian diaspora in the Gulf countries faces a range of challenges that impact their daily lives and overall well-being. These challenges include labour rights issues, substandard living conditions, social integration difficulties, and limited access to legal and consular support. Addressing these issues is crucial for improving the quality of life for Indian expatriates and ensuring their contributions to the Gulf's economy and society are recognized and valued. Looking ahead, there are promising opportunities for the Indian diaspora as Gulf countries diversify their economies and strengthen bilateral relations with India. Enhanced economic prospects, cultural diplomacy, policy reforms, and technological advancements offer a brighter future for Indian expatriates, fostering a more inclusive and supportive environment.

Labour Rights and Exploitation: One of the most significant challenges faced by Indian workers in the Gulf is the exploitation under the Kafala (sponsorship) system. This system ties workers' legal status to their employers, often leading to situations where workers face harsh working conditions, wage theft, and limited freedom to change jobs. Despite recent labour reforms in some Gulf countries, enforcement remains inconsistent, and many workers still experience exploitation and abuse.

Living Conditions: Many Indian workers, especially those in low-wage jobs, live in overcrowded and substandard accommodations. These conditions were highlighted during the COVID-19 pandemic, which saw high infection rates among migrant workers due to cramped living spaces. Ensuring better living conditions and access to healthcare remains a critical issue.

Social Integration: Indian expatriates often face challenges in integrating into Gulf societies due to cultural differences and social barriers. While there are vibrant Indian communities and cultural organizations, the broader integration into the host society can be limited. This can lead to feelings of isolation and a lack of social support networks.

Legal and Consular Support: Access to legal and consular support can be limited, particularly for low-wage workers. Many workers are unaware of their rights or lack the resources to seek legal assistance when facing issues such as contract violations or abuse. Strengthening legal support and awareness programs is essential to protect the rights of Indian workers.

Economic Diversification and Opportunities: As Gulf countries diversify their economies away from oil dependence, new opportunities are emerging for skilled Indian professionals in sectors such as technology, healthcare, and education. This shift can lead to better job prospects and improved working conditions for Indian expatriates.

Enhanced Bilateral Relations: The Indian government has been proactive in strengthening bilateral relations with Gulf countries, focusing on labour agreements, trade, and investment. Continued diplomatic efforts can lead to better protection for Indian workers and increased economic collaboration, benefiting both regions.

Cultural Diplomacy and Soft Power: The Indian diaspora's active engagement in cultural and social activities enhances India's soft power in the Gulf region. Promoting cultural exchanges and fostering people-to-people connections can improve mutual understanding and integration. This cultural diplomacy can also create a more inclusive environment for Indian expatriates.

Policy Reforms and Advocacy: Ongoing advocacy efforts by Indian community organizations and diplomatic missions are crucial in pushing for policy reforms that protect migrant workers' rights. Strengthening these efforts and ensuring effective implementation of labour laws can significantly improve the living and working conditions of Indian expatriates.

Technological Advancements: The adoption of digital platforms and technologies can facilitate better communication and support for Indian workers in the Gulf. Online portals for legal assistance, health services, and community engagement can provide much-needed resources and support. By addressing these challenges and leveraging future opportunities, the Indian diaspora can continue to play a vital role in the development of the Gulf region while enhancing their own quality of life.

Conclusion

The Indian diaspora in the Gulf countries stands as a testament to the profound impact of migration on both host and home nations. Their contributions to the Gulf's economy and society are multifaceted, encompassing labour, entrepreneurship, cultural exchange, and diplomatic engagement. The economic interdependence forged through remittances and labour has not only bolstered India's economy but also fueled the rapid development of Gulf nations, transforming their urban landscapes and industrial capacities. Culturally, the Indian community has enriched the Gulf's social fabric, fostering a vibrant multicultural environment that celebrates diversity and promotes mutual understanding. Through festivals, community events, and cultural organizations, the diaspora has preserved its heritage while also contributing to the cultural dynamism of the Gulf region. This cultural diplomacy enhances India's soft power, creating a positive image and fostering goodwill. Politically, the Indian diaspora serves as a crucial link between India and the Gulf countries, facilitating bilateral relations and advocating for the rights and welfare of expatriates. The proactive engagement of the Indian government and community organizations in addressing labour rights issues and improving living conditions underscores the importance of this diaspora in shaping policy and diplomatic efforts. Looking ahead, the future of the Indian diaspora in the Gulf is filled with promise. As Gulf countries diversify their economies and embrace technological advancements, new opportunities will emerge for skilled Indian professionals, further integrating the diaspora into the socio-economic fabric of the region. Enhanced bilateral relations, driven by strategic partnerships and cultural diplomacy, will continue to strengthen the ties between India and the Gulf, fostering mutual development and regional stability. In conclusion, the Indian diaspora's contributions to the Gulf's economy and society are invaluable. By addressing the challenges they face and leveraging their potential, this community can continue to be a driving force for development and a bridge of friendship between India and the Gulf countries. Their story is one of resilience, adaptability, and enduring impact, shaping the present and future of both regions in profound ways.

References

Abraham, R. (2012). India and its diaspora in the Arab gulf countries: Tapping into effective'soft power'and related public diplomacy. Diaspora Studies, 5(2), 124.

Anjum, F. (2017). Indian Diaspora in GCC Countries. *International Journal of Research Culture Society*. 1(10).

Auwal, M. A. (2010). Ending the exploitation of migrant workers in the Gulf. Fletcher F. World Aff., 34, 87.

Bhandare 2020- https://www.thehindu.com/opinion/columns/when-the-rupee-ruled-mughal-currency-trade-coin-east-india-company/article68303999.ece.

Biswas, A. (2021). Changing dynamics of the Indian diaspora in the Gulf Region. In New Perspectives on the Indian Diaspora.

Cohen, S. I. (2008). Dualistic Development in UAE and GCC. *Topics in Middle Eastern and African Economies*. 10.

Faheem M, Azali, M, Chin, L. & Mazlan, N. S. (2022). Migrant Remittances and Dutch Disease: Evidence from India. Iranian Economic Review, 26(1): 121–132.

Fawcett, J. T. (1989). Networks, Linkages, and Migration Systems. *The International Migration Review*, *23*(3), 671–680. https://doi.org/10.2307/2546434.

Fouron, G., & Schiller, N. G. (2001). All in the Family: Gender, Transnational Migration, and the Nation-State. Identities, 7(4), 539–582. https://doi.org/10.1080/1070289X.2001.9962678.

Gardner, A. M. (2011). Gulf migration and the family. Journal of Arabian Studies, 1(1), 3–25.

Jain, P. C. (2005). Indian Migration to the Gulf Countries: Past and Present. *India Quarterly*, *61*(2), 50–81. http://www.jstor.org/stable/45073330.

Jones, R. (2016). Beyond Kafala: remedying human rights abuses of migrant workers in the Persian Gulf.

Kritz, M.M., Lim L.L. and Zlotnik, H. (1992) International Migration Systems: A Global Approach. Clarendon Press, Oxford University Press, Oxford.

Mabogunje, A. L. (1970). Systems Approach to a Theory of Rural-Urban Migration. *An International Journal of Theoretical Geography*. 2(1), 1–18.

Naha, A (2024). The Dynamic Role of the Indian Diaspora in Shaping India's Relationship with the Middle East. In Anshu A. (Ed.) Diaspora in the MENA region and beyond. UJ press: Johannesburg.

Nayudu, T. (2022). *The Indian Diaspora in the Gulf States* [Unpublished master's thesis]. Masaryk University.

Onley, J. (2014). Indian Communities in the Persian Gulf, c. 1500–1947. In: Potter, L.G. (eds) The Persian Gulf in Modern Times. Palgrave Macmillan, New York. https://doi.org/10.1057/9781137485779_10.

Rasool, F., & Haider, S. I. (2020). Exploitation and Low Wages of Labor Migrants in Gulf Countries. Global Management Sciences Review, 1, 32–39.
Singh and Arimbra. (2019). https://medium.com/@indiamigration/indians-in-the-gulf-the-other-side-of-the-story-2870995eb748.
Upegui-Hernandez, D. (2014). Transnational Migration Theory.
Weiner, M. (1982). International Migration and Development: Indians in the Persian Gulf. *Population and Development Review*, *8*(1), 1–36. https://doi.org/10.2307/1972688.

Diaspora and Development: A Study of Indian Diaspora in South Africa

Santosh Kumar Singh

Introduction

The India and South Africa relationship has strengthened its collaboration with since after the Mahatma Gandhi's struggle against the apartheid government in South Africa. This relationship is based on values, trust, mutual support and experience. As it has been mentioned by the first President of South Africa Nelson Mandela 'India and South Africa are two countries held so closely by bonds of sentiment, common values and shared experience, by affinity of cultures and traditions and by geography' (Tripathi 2023). Even though, since independence (1947), India has supported and fought against the matter of apartheid on this subject at United Nations, Non-Alignment Movement (NAM) and many more multilateral and global organisations. Along with the political ties both the nations established diplomatic ties, the trade has increased from US Dollar 10 million in 2015–2016 to US Dollar 19 million in 2022–2023 (Consulate General of India 2025) (Table 1).

S. K. Singh (✉)
Department of Political Science, Aryabhatta College, University of Delhi, New Delhi, Delhi, India
e-mail: santosh1singh@gmail.com

A. Dubey and S. Rajwar (eds.), *Diaspora in Development*,
https://doi.org/10.1007/978-981-96-6747-5_7

Table 1 India–South Africa bilateral trade (Figures in Million USD $)

	2018–2019	*2019–2020*	*2020–2021*	*2021–2022*	*2022–2023*	*2023–2024**
India's exports	4,067.20	4,108.17	8,934.20	6,085.29	8,474.42	6,263.23
India's imports	6,517.33	6,969.79	7,568.50	10,965.81	10,397.83	8,266.95
Total:	10,584.53	11,077.96	16,502.70	17,051.10	18,872.25	14,530.18

* Apr.–Dec. 2023

Source Consulate general of India (2025), India—South Africa bilateral economic engagement, accessed from https://www.cgijoburg.gov.in/india-south-africa-bilateral-economic-engagement.php on 12th January 2025

Moreover, in the era of liberalisation and globalisation the diaspora has evolved as powerful entity. It has been recognised as a 'soft power' in the realm to establishing the diplomatic and strategic relationship between the host and home nation. Diaspora is also working as an agent of economic cooperation between the home country—country of origin long with their active role in the host countries—country of residence, for instance, as it can be seen in the case of Chinese Diaspora. The Chinese Diaspora has been recognised as a driving force in strengthening its economic relations and negotiations with host nation. In other instances it can be seen in the context of Jewish Diaspora. Jewish Diaspora has strong influence over the United States of America's relationship and European Countries in terms of shaping their strategic relationship with Israel (Khara 2020). In similar condition around 1 million people constitutes that is around 3% of the population constitutes the South African Indian origin people. Indian diaspora is also playing a pivotal role to strengthen the relationship between India and South Africa from various political, social and economic perspectives.

Diaspora as an Actor

During the twentieth century, the term 'diaspora' is represented as a dispersal of any ethnic community or group of people outside the country of their birth. In contemporary era, it is recognised in most part of the world such as Greek diaspora, Indian Diaspora, Japanese Diaspora, Italian Diaspora, Jewish Diaspora, Chinese Diaspora, Lebanese Diaspora, South Korean Diaspora and Polish Diaspora. Since the end of the cold war, the Diaspora has played pivotal role to establish the relationship between the nations. It has contributed to their home as well as the host nation through remittances, philanthropy, skill transfer, investment, advocacy and assistance.

India holds world's second-largest diaspora, with an expansive overseas Indian community exceeding 35.42 million, strategically embedded across key global regions. It includes around 15.85 million non-resident Indians (NRIs) and around 19.57 million individuals of Indian origin (PIOs) (Indian Express 2025). The Indian Diaspora has evolved through successive era of migration over centuries, driven by factors such as mercantilism, colonialism and globalisation. Its early journey is a testament to resilience, marked by hardships, perseverance and eventual success. In the last three decades of twentieth century, the nature of movement of people underwent a significant shift, giving rise to a 'new diaspora'—comprising highly skilled professionals migrating towards the Western country and semi-skilled contract workers seeking opportunities in the West Asia and Southeast Asia (MOIA 2015).

In the end of twentieth century India has witnessed a paradigm shift in its diaspora population. India is trying to establish good fortune and closer links with its overseas community. In 2000, the Government of India has established the first-time setup High Level Committee on the subject 'Indian Diaspora' under the chairmanship of Dr. L. M. Singhvi. He employed 'diaspora' as a generic sense for '...communities of migrants living or settled permanently in other countries, aware of its origins and identity and maintaining varying degrees of linkages with mother country' (Mukhopadhyay 2010).

The Indian government institutionalised the engagement with Indian diaspora through the establishment of the Ministry of Overseas Indian Affairs (MOIA) in 2004, which later kept inside the Ministry of External Affairs (MEA) in 2016, to streamline diaspora-related initiatives (MEA 2015). Furthermore, initiatives like Pravasi Bharatiya Divas (PBD), launched in 2003, celebrate the achievements of the Indian diaspora and foster dialogue between Non-Resident Indians (NRIs) and policymakers. On January 9, the Government of India celebrates PBD to honour the contributions of overseas Indian community in the country's development. This date marks the return of Mahatma Gandhi to India from South Africa in 1915, a pivotal moment that shaped India's struggle for independence. Additionally, the Overseas Citizenship of India (OCI) scheme, introduced in 2005, provides lifelong visas and other privileges to the diaspora, enhancing their connection to the homeland (Government of India 2005).

Diaspora Policy During Nehruvian Era

Historically, if we look back to the India's diaspora policy it can be start from the first Indian Prime Minister Pandit Jawaharlal Nehru who has seen it as somewhat detached from India. Furthermore, he mentioned to make their own destinies in their host countries. Nehru categorically mentioned about the Indians who left their country (home country) seeks employment at global level had to integrate with the host country. He mentioned 'It is the consistent policy of the Government that persons of Indian origin who have taken foreign nationality should identify themselves with and integrate in the mainstream of social and political life of the country of their domicile. The government remains naturally alive to their interests and general welfare and encourages cultural contacts with them. As far as Indian citizens residing abroad are concerned, they are the responsibility of the Government of India' (Lall 2001:169).

India has adopted noninterventionist policy regarding the India Diaspora. In other words, it has been mentioned as policy of *active dissociation*. According to Lall (2001), this approach had several limitations. During this period, many members of the Indian Diaspora were enduring significant political, economic and social crises. India, however, was reluctant to engage with these issues, as it feared damaging its relations with the newly decolonized countries. Consequently, the Indian government refrained from addressing human rights violations affecting its diaspora, prioritizing diplomatic considerations over the concerns of its overseas population (Gangopadhyay 2005).

In 1990s India has liberalized its economy and adopted *new economic policy* it has led the immediate and extensive economic and political reforms. This policy forced India to open its economy and market for foreign investors. It has also given way to Indian government to constitute a committee under the department of industrial development and for the industrial proposals of NRIs (Non-Resident Indians). Thus, on the basis of this, in 1987 the Indo-NFU Chamber of Commerce and Culture was established to encourage the overseas Indians' cases. According to Parekh (1993), the primary objective behind this initiative was, first and foremost, to challenge long-standing stereotypes as increased engagement between Indians and the overseas expatriate community took place.

This interaction instilled a profound sense of pride in their resilience and achievements, fostered a commitment to recognizing their contributions, and highlighted the prolonged neglect they had endured (Mahajani 1976). Secondly, India's strategic self-interest was also a key factor. With a deteriorating foreign exchange situation since the 1970s, the nation recognised the success and prosperity of Indians who had migrated to developed countries post-independence. India sought to leverage not only their financial remittances but also their expertise in technology, science, management and other critical fields (Parekh 1993).

The Indian government subsequently signed multiple treaties with various West Asian countries on the subject such as deportation, labour laws and amendments to its own in 1983 Emigration Act. Additionally, during the Middle East war, the Special Coordination Unit and the Overseas Indians Division played a crucial role in the evacuation of Indian workers from the region (Parekh 1993). Several special concessions were introduced to encourage Indian diaspora investment in Indian industries, facilitate new industrial ventures and attract foreign currency deposits in Indian banks. The key incentives included permission for Indian diaspora investment in real estate development, invest in India Development Bonds and the automatic approval of investments and technical collaborations (Ibid).

The 'New' Diaspora Policy Under BJP Government

It was the Bhartiya Janta Party (BJP) government, who brought a transformative shift in the diaspora policy. They have taken several reforms measures towards to establish the links with Indian Diaspora. This shift was clearly visible Chennai Declaration which was held in 1999 when the BJP mentioned;

> We believe that the vast community of NRIs and PIOs also constitute a part of the Great Indian Family. We should endeavour to continually strengthen their social, cultural, economic and emotional ties with their mother country. They are the rich reservoir of intellectual, managerial and entrepreneurial resources. The Government should devise innovative schemes to facilitate the investment of these resources for India's all-around development. (Lall 2001:98)

Along with this Chennai Declaration, the government has taken several measures in such regard government has also proposed an announcement for NRIs to launch the People of Indian Origin Card Scheme on 31 March 1999. The PI0 card allowed for some special economic, educational, financial and cultural benefits besides acting as a long-term visa for the cost of $1000 for the duration of 20 years (Lall 2001).

The High-level committee on Indian diaspora under L.M. Singhvi emphasised the need of clear and defined policy to establish the country-specific plans for enhancing connectivity. It has been suggested to improve the airports, rules and regulations the government, solve the problems of overseas Indian labour. There are sector-wise recommendations too under the headings of Culture, Economic Development, Tourism, Education, Health, Media etc. During the first Pravasi Bharatiya Diwas (PBD), the former Prime Minister Atal Behari Vajpayee mentioned.

> We are in favour of dual citizenship but not dual loyalty. The loyalty with India will remain but they will also be loyal to the country where they have taken citizenship but it has been resolved now. I am hopeful that Indians settled abroad will find it suitable. (Bharti 2003)

One of the key changes initiated by the BJP government in 2014–2015 is to reaching out to the Indian Diaspora with the help foreign policy initiatives. This was evident from his interactions with the diaspora communities in the US, Australia and Canada. On his foreign visits so far Prime Minister Shri Narendra Modi addressed the diaspora in jam-packed events. Through such events the government tried to establish contact with the diasporic community directly. During these events, the government played twin roles, on one hand it has sought the assistance in transforming the country, not just by investing, but also contributing. The Indian government also initiated single identity-based card—easily connect the Indian diaspora. In 2014 the Indian government has also initiated right to vote for Indian citizens living abroad to vote in Indian elections(Rama 2015).

India's Engagement with South Africa

In terms of numbers, South Africa has the largest presence of Indian communities on the African continent (Renu 2010). The Indians in Africa are categorised under the two main categories, the 'old diaspora' and the 'new diaspora'. The old diaspora group represents the descendants of indentured labourers also known as *'girmitiyas/slaves'* who were dispatched from India during 1860s to 1911. Most of the Indian people have taken from the Gujarat, Maharashtra, Punjab (of pre-partition India), Bengal (including Bihar and Orissa), Tamil Nadu and Kerala. After the abolition of slave trade most of the Indians were taken to South Africa especially at the Natal province as a free passengers and traders (Mesthrie 2000).

In 1946, the Asiatic Land Tenure Act that is also known as *Ghetto Act*, restricted the Indians to residential areas reserved for them. As a result, Indian communities were resettled in large residential townships like Chatsworth and Phoenix in Natal, which were specifically developed for them. The apartheid system, which was established in 1948, performed a key role to influence the development of Indian diaspora. This system has established the discriminated laws and practices against the non-white South Africans. It has also applied on the Indians in South Africa. It has compelled many Indians to establish their own groups, forums and associations. This discrimination and suppression have paved the way to the established newly 'Indian' townships (Pachai 1971).

Apart from this, India has worked consistently at the international platforms against the apartheid regime in South Africa either it was United Nations (UN), Non-Alignment Members (NAM) and various international platforms. On the other hand, since 1960s South Africa with the help of African National Congress (ANC) has maintained a representative in India to sustain struggle (Ministry of External Affairs 2024). The role of Indian diaspora to establish multiple ties with South Africa can be seen under different phases and era.

Relations in the Post-Apartheid Era

India and South Africa have been witnessed of the oppressive colonialism. In post-independence period one of the major objectives in Indian foreign policy was working for decolonisation and supporting anti-colonial struggle in the various parts of the world. Since diplomatic relations were established in November 1993, India and South Africa have signed numerous agreements and MoUs across various sectors, including political, economic, defence and cultural cooperation. On 9th May 1993 after a gap of over four decades, diplomatic relations between India and South Africa were formally restored after the opening of a Cultural

Centre at Johannesburg. Along with this, in 1994 India opened its High Commission at Pretoria opened the Consulate General in Durban (Ministry of External Affairs 2024). Since then, the evolving transnational reconfigurations of South African Indian identity have emerged from a convergence of factors aimed at strengthening ties with India. These include the Indian government's renewed commitment to engaging the South African Indian community as part of its broader diaspora strategy, the expanding role of South African Indian entrepreneurs in fostering economic connections, and the growing influence of Bollywood promoters and film producers from both India and South Africa in shaping cultural exchanges (Dickinson 2014).

The contemporary partnership between these two countries derive their sustenance from the strategic partnership agreement signed between the two countries. The initial strategic partnership initiated by the Deputy President Mbeki during his visit to India in 1996. It was in 1997, the strategic partnership was declared from the Red Fort in New Delhi during the President Nelson Mandela's visit to India known as the *Red Fort Declaration*. This declaration has enhanced the bilateral cooperation and collaboration between them especially at international forums (Robin 1997) and (Edgars 2019).

Apart from this, the Indian High Commission has also been opened in Pretoria in May 1994. It was inaugurated by President Nelson Mandela. In post-apartheid era, it has been found that Indian Diaspora in South African has represented in the government, business, media, legal and other professions. It has played an important role to establish the multiple ties between India and South Africa. In the continuation of this, the year 2010 has been celebrated on the occasion of 150th anniversary of the first arrival of Indians in South Africa. The year 2014 has been marked as 100 years of Gandhi's return to India from South Africa. In 2023, India celebrated 30 years of diplomatic relationship (Financial Express 2023).

In South Africa many South African Indian belongs to various political parties, even they have won back-to-back elections at National Assembly and provincial legislatures, for example Mergan Chetty won his third consecutive stint national assembly in 2024. Similarly, Shara Singh, member of Provincial Legislature from KwaZulu-Natal, Ashor Saruppen won his second straight National Assembly, Narend Singh, leader of Inkatha Freedom Party, Fasiha Hassan of African National Congress (ANC), Anilkumar Kesava Pillai and many more (Sudhi 2024). They are playing a pivotal role to establish the diplomatic and strategic ties with India based on the model of mutual development.

Diaspora Networks and Development Partnerships

India and South Africa have very close cultural ties. It is not only about the numbers of South African Indian origin community which constitutes around 3% (around 1.5 million) of South Africa's total population but it is about the culture and practice which they are still carrying and practising—the language, religion, customs, festivals and living style. In South Africa. According to Consulate General of India (Durban), in 2025, around 80 percent of Indian people resides in KwaZulu-Natal province and around 15 percent live in Gauteng province and the remaining residents in Cape Town (5 percent). The most important contribution by Indian Diaspora in South Africa is they are representing and contributing their roles at all the spheres such as business, mass-media, government, legal and other professional areas (Ministry of External Affairs 2015).

According to the Ministry of External Affairs (2025a), in South Africa, India has established several institutions to strengthen its multifaceted ties with the Indian diaspora. India has established one High Commission of India (HCI) in Pretoria and three CGIs in Durban, Johannesburg and Cape Town. The CGI Johannesburg was established in 1993 during the end of apartheid regime. On May 1994, another CGI in Durban was established. HCI was founded in 1994 at Pretoria and third CGI established in 1996 at Cape Town. These Consulates provided services to local, Indian and international citizens in South Africa. It includes the general or specific information regarding to obtain Indian Visa, Passport services and other issues. Along with this in 1994 the Indian Cultural Centre (ICC) established in Durban and Johannesburg ICC (established 2011)—all these institutions founded to respond the increasing role in India's soft power in South Africa (Dickinson 2014:81).

The CGI, Durban in collaboration with Indian Council for Cultural Relations (ICCR) and local Indian cultural associations arranges and supports to provide platform and arrangement for their activities—Indian artists and language education. Since 2004, the ICC Durban has organised various initiatives with the help of the Ministry of Overseas Indian Affairs (MOIA). It is performing a major development role to promote and support Indian language and cultural education programming. In this regard, the ICC also performing important role to collaborate with local

Indian educational, linguistic and Hindu cultural diaspora associations. It includes wider range of local Indians who are performing art schools, business-people and key opinion formers (Dickinson 2014(a)). ICCR has organised several programs, events and activities to support the various delegations of South African Indian artists and performers (Dickinsion 2015).

Moreover, to organise and develop the links and contribution to Indian Diaspora the government of India has taken initiative to reward India through awards the Indian community people for their excellent work in the fields of business, philanthropy, arts and social service (Ministry of Overseas Indian Affairs 2009). In this regard, in 2010 Dr. TP. Naidoo, who had established the Indian Academy of South Africa in [;] was awarded a *Pravasi Samman Award* for his exemplary work to promote Indian culture in South Africa (Ministry of Overseas Indian Affairs 2009). In a press interview TP. Naidoo mentioned.

> an awe-inspiring experience to receive this award from the head of the largest democracy in the world [...] The award is held in great esteem and wherever I went after receiving the award, people treated me with dignity and congratulated me. The media had splashed the news overall major networks on radio and TV. I dedicate this award to the South African Indian community who despite tremendous trials and tribulations continued to triumph against all odds to hold their own on the world stage. (Artsmart 2010)

The Indian Government has taken several initiatives to strengthen the ties with Indian Diaspora in South Africa, such as in 2002 the Indian High Commission in South Africa has taken initiative to repair Gandhi's ashrams in South Africa. The Madhya Pradesh government has also taken initiative to spend 1 crore ($215,000 USD) to support the Indian High Commissions' efforts to renovate the Tolstoy Farm with the help of South African India (The Hindu 2014). On this occasion the Indian High Commissioner Virendra Gupta mentioned 'To commemorate 100 years of Gandhi's return to India from South Africa in 1914, a function is being organised by the High Commission of India in collaboration with the local Indian community and the Government of South Africa on 20 July 2014 at the Tolstoy Farm, (Ibid)'. Such initiative has not only strengthened the ties with Indian diaspora in South Africa but also strengthened the relations between the countries.

Trade and Economic Cooperation

India and South Africa have old traditional trade relationship. This trade ties have flourished immensely after the post-apartheid era. In this regard, the South African Indian have performed a role as a bridge to facilitate the investment and business collaborations. According to the Indian Ministry of Commerce (2025), bilateral trade between India and South Africa grew from approximately $2 billion in 1999 to over $20 billion by 2025. The value of bilateral trade has increased from US$2.5 billion in 2003–2004 to US$19.25 billion in 2023–2024. In this trade South Africa exports the coal, copper, manganese ore, gold and bulk minerals to India. Meanwhile, India's major exports to South Africa include petroleum products, vehicles, pharmaceuticals, telecom equipment, textiles, agricultural products and construction machinery products (Department of Trade, Industry and Competition 2023) (Table 2).

In this trade and business relations, the presence of deeply entrenched Indian community, particularly in Durban, Johannesburg and Cape town has facilitated the trade ties, business networks, promoting investment and serving as cultural intermediaries in negotiations. Along with this, the Confederation of Indian Industry (CII) and such other Indian-origin business chambers in South Africa are also performing incredible role in market expansion, advocacy and fostering partnerships. It is the role of India as well as the Indian community contribution in the development of Small and Medium-sized Enterprises (SMEs), which serve as an essential pillar of economic engagement between both the nations (Bhana and Brain, 1990). India's Micro, Small and Medium Enterprises (MSMEs) sector contributes 45 percent to the South Africa's manufacturing output and 40 percent of the total exports. Along with this it also provides employment over 60 million people (The Economic

Table 2 India-South Africa trade (US $ Millions)

Year	*2019–2020*	*2020–2021*	*2021–2022*	*2022–2023*	*2023–2024*
Export	4,108.17	3,934.20	6,085.29	8,474.42	8,707.53
Import	6,969.79	7,568.20	10,965.81	10,397.83	10,538.23
Total trade	11,077.97	11,502.39	17,051.10	18,872.24	19,245.77

Source Ministry of Commerce (2025), Ministry of Commerce: Export import data bank, accessed from https://tradestat.commerce.gov.in/eidb/iecnt.asp on 17th January 2025

Times 2012). In South Africa, Small, Medium and Micro Enterprises (SMMEs) contribute around 91 percent of businesses and employees around 60 percent labour force. It has also contributed approximately 34 percent economic production (International Finance Corporation 2018) (Campbell 2013). Account to advisory director and head of India Abdullah Verachia pointed out that there are now 110 substantial Indian companies doing business in South Africa. Furthermore, he said.

> There is a very rapidly growing economic and commercial relationship [between India and South Africa]. Hardly a week passes by without us hearing about a new investment decision, either by a South African company in India or an Indian company in South Africa. The relationship has not reached a plateau (Campbell 2013)

Indian companies such as Tata, Mahindra and Vedanta have made significant investments in South Africa's mining, automobile and energy sectors, generating employment and fostering industrial growth (Mukherjee 2021). Vedanta Resources chairperson Anil Agarwal mentioned that the Vedanta's investment in South Africa is majorly due to its sentimental attachment, along with this it also holds landmark point of India's father of nation Mahatma Gandhi's freedom struggle land. It was in 2018, the chairperson mentioned to investing $1-billion in South Africa, with the aim to create substantial number of jobs (Parker 2023).

Former President of South Africa, Nelson Mandela, once said, 'India and South Africa are two countries held so closely by bonds of sentiment, common values and shared experience, by the affinity of cultures and traditions and by geography'. This can be demonstrated through, for example, the way that the Durban mayor, Logie Naidoo (a South African Indian) articulated the benefits of Indian diasporic sensibilities. He said:

> On my last visit to Chennai, I attracted Ramco, an IT company, to Durban and they invested [...] It's important I maintain links to India to assist the South African community [to] provide a platform of mutual benefit to both countries. I use my identity, origins and position to work on the links between India and South Africa and build links and extract benefits for South Africa. (Dickinson 2015)

Pravin Gordhan, a highly esteemed politician of Indian origin, was appointed as South Africa's new Finance Minister by President Jacob Zuma. His appointment marked the third change in the country's finance leadership within a week, occurring amidst a period of intense economic turmoil (Times of India 2015). According to data from the United Nations Comtrade database, South Africa's imports of pharmaceutical products from India amounted to US$599.91 million in 2023. This data represents India's critical role as a key supplier of affordable and essential medicines to South Africa. India, known as the 'pharmacy of the world', has been a major source of generic medicines, including life-saving drugs such as antiretrovirals (ARVs) used in South Africa's national HIV/AIDS treatment programs. The strong pharmaceutical trade relationship between the two countries is driven by India's ability to produce high-quality, cost-effective medicines, which help improve healthcare accessibility in South Africa.

India plays a crucial role as a key supplier of affordable essential drugs to South Africa. This partnership on medicine access dates back to 2001, when the Indian generic drug manufacturer Cipla offered to sell the generic antiretroviral (ARV) drugs for less than a dollar per day. At present, the South African government relies heavily on Indian made drugs for various of its national treatment programs especially HIV/AIDS. During the time of pandemic COVID-19, India supported South Africa through providing additional vaccines, research cooperation and medical equipment (The Wire (2021). On 28th June 2021, South Africa Minister for International Relations and Cooperation Naledi Pandor met with India's Foreign AM Dr Jaishankar on the event of G20 Foreign Ministers Meet in Italy, their they mentioned that both countries are working together to meet the COVID-19 challenge and about economic recovery and Commonwealth issues (Government of India 2024).

Pravasi Bharatiya Divas (PBD)

In 2025, India has celebrated 18th Pravasi Bharatiya Divas (PBD) to endorse the contribution of Indian Diaspora in the development of India. On this event more than 3,000 delegates from across more than 70 countries focused on mutual self-help (The Hindu 2025). Many delegates from South Africa have also participated in this occasion. South Africa's High Commissioner to India Anil Sooklal said, it was a memorable and privilege to attend the Pravasi Bharatiya Divas (PBD) in Odisha's Bhubaneswar. There he recalled the memories of PBD, which was held in Varanasi in 2019. He said.

> I think today the Indian diaspora is an integral part of India's architecture. You cannot speak of India without the Indian diaspora. It is two sides of the same coin. As you see the rise of India, you also see the prominence of the global Indian diaspora.........A number of, I always speak of a Global South diaspora because I come from South Africa, a country from the Global South. And those of us who were part of the indenture system, we're very proud that we have our culture embedded in the Indian subcontinent and the PBD is a recognition of our identity, our heritage, and our roots in many senses. (ANI 2025)

To Stregthen the relations with Indian Diaspora , the Indian government has taken the initiatives to organise PBD regionally—to connect Indian Diaspora who are not able to participate in Pravasi Bharatiya Divas in India. Untill now total of ten Regional Pravasi Bharatiya Divas (RPBD) has been organised at New York, Singapore, The Hague, Durban, Toronto, Mauritius, Sydney and London. The fourth RPBD organised in the series was held in Durban. That was themed as 'PBD Africa' held on (1st-2nd October) 2010. On this occasion the South Africa President Mr. Jacob Zuma graced the closing session on 2nd October 2010. This event was also marked as the 150th anniversary of the arrival of Indians in South Africa (Ministry of External Affairs 2025a). On this occasion, he exhorted people of India Origin (PIO) to strengthen economic ties: 'For Africans of Indian origin, the bonds tying them to the land of their origin have hitherto been social and cultural. In the changing economic paradigm, you must increasingly look at India as a land of economic opportunity' (Ministry of Overseas Indian Affairs 2010a).

Conclusion

Indian culture has played a pivotal role to build and establish the past, present and will continue to do so in the future as well. As we can see festivals like Holi, Diwali, Shivratri, Durga Puja and many more cultural festivals celebrated by great enthusiasm in Mauritius, Fiji, Guyana, Trinidad and South Africa. As it has been mentioned by Joseph Nye that in the era of modernity Diaspora is working as an important actor to establish the multiple ties between the host and home nation. Especially after the information technology revolution and better transport linkages eliminated the geographical and information-based gap between the people and information. As it is evident through several policy initiatives taken between India and South Africa, such as double tax avoidance treaty, COVID-19 vaccine, trade ties etc. The Diaspora is playing direct and indirect role to strengthen the economic and political but also cultural ties.

The Indian community in South Africa occupies diverse economic, political and social positions. Economically, they are a highly influential and prosperous group, with the majority engaged in business and trade rather than formal employment. Predominantly belonging to the business class, they have established themselves as a key force in South Africa's commercial landscape. Their entrepreneurial ventures, ranging from small enterprises to large-scale industries, have solidified their role as the largest trading community, wielding significant yet discreet influence over the country's economic dynamics. Their contributions extend beyond commerce, shaping the broader economic framework of South Africa. With over 1.5 million South Africans of Indian descent, their influence is evident in various sectors, including business and politics.

References

ANI (2025), "Great privilege to be part of Pravasi Bharatiya Divas": South African envoy Anil Sooklal,' Accessed from https://thenewsmill.com/2025/01/great-privilege-to-be-part-of-pravasi-bharatiya-divas-south-african-envoy-anil-sooklal/.

Artsmart, (2010), 'Dr Tp Naidoo Honoured in India', accessed from http://news.artsmart.co.za/2010/02/dr-tp-naidoo-in-india.html on 17th May 2018.

Bhana, S., Brain, J.B., (1990), *Setting down Roots: Indian Migrants in South Africa, 1860–1911*, Wits University Press, Johannesburg.

Bharti, Vivek. (2003), 'Pravasi Bhartiya Divas'. Think India.Vol.6, No.I, pp.18–19.

Bhiku Parekh (1993), "The Indian Diaspora", in Jagat K. Motwani, Mahin Cosine and Jyoti Barot Motwani, eds., The Global Indian Diaspora: Yesterday, Today and Tomorrow, New York: Global Organization of People of Indian Origin, pp. 10.

Campbell, Keith (2013), 'Economic Relations Between South Africa and India Blossoming', accessed from https://www.engineeringnews.co.za/print-version/economic-relationship-between-south-africa-and-india-is-blossoming-2013-05-10 on 15th May 2019.

Cohen Robin (1996), "Diasporas and the Nation-State: From Victims to the Challengers" *International Affairs*, 72(3): 507–20.
Cohen Robin, 'Global Diasporas: An Introduction' (London: UCL Press 1997)
Consulate General of India (2025), 'India - South Africa Bilateral Economic Engagement'', accessed from https://www.cgijoburg.gov.in/india-south-africa-bilateral-economic-engagement.php on 12th January 2025.
Dickinson, J (2014) (a), Making Space for India in Post-Apartheid South Africa: Nnarrating Diasporic Subjectivities Through Classical Song and Dance, *Emotion, Space and Society*, 13:32–39, accessed from https://doi.org/10.1016/j.emospa.2014.05.001.
Dickinson, Jen (2015), Articulating an Indian diaspora in South Africa: The Consulate General of India, diaspora associations and practices of collaboration, *Geoforum,* 61: 79–89, p. 85.
Dupelia-Mesthrie, U., (2000), From Cane Fields to Freedom: A Chronicle of Indian South African Life, Kwela Books: Cape Town.
Edgars, Tal (2019), 'India—South Africa Economic Relations: The Red Fort Declaration,' The Diplomatist, accessed from https://diplomatist.com/2019/12/15/india-south-africa-economic-relations-the-red-fort-declaration/ on 12th May 2021.
Financial Express (11 January 2023), 30 years of India-South Africa relations: Connected with Cricket, Culture, Cuisine, accessed from https://www.financialexpress.com/life/travel-tourism-30-years-of-india-south-africa-relations-connected-with-cricket-culture-cuisine-2943926/ on 12th January 2025.
Gangopadhyay, Aparajita, India's Policy towards its Diaspora: Continuity and Change, *India Quarterly: A Journal of International Affairs*, 2005, vol. 61(4):93–122.
Gilbert, H., Lo, J., (2010), Diasporas and performance. In: Knott, K., McLoughlin, S.(Eds.), Diasporas: Concepts, Intersections, Identities. Zed, London, pp. 151–156.
Ginwala, F., (1985), *Indian South Africans*, MRG, London, P-3.
Government of India (2024), India-South Africa Economic Relations: A Comprehensive Report. Ministry of External Affairs, accessed from https://www.mea.gov.in/Portal/ForeignRelation/Bilateral-brief_SA-for-XPD-June-2024.pdf
Indian Express, 2025, Where do most Indians live abroad? Top 10 countries with the largest Indian diaspora, 9th January 2025 accessed from https://indianexpress.com/article/trending/top-10-listing/top-10-countries-with-the-largest-indian-diaspora-2024-9727289/ on 24th January 2025.
International Finance Corporation (2018), 'The Unseen Sector: A Report on the MSME Opportunity in South Africa,' accessed from https://documents1.worldbank.org/curated/ar/710801548830275900/pdf/134151-WP-ZA-Unseen-Sector-MSME-Opportunity-South-Africa-PUBLIC.pdf.

Jen Dickinson 2014), 'Making space for India in post-apartheid South Africa: Narrating diasporic subjectivities through classical song and dance', *Emotion, Space and Society*, 13:32–39.

Khan, Sultan (2012), 'Changing Family Forms, Pattems and Emerging Challenges Within the South African Indian Diaspora', Journal of Comparative Family Studies, Vol. 43(1):133–150, p. 134.

Khara, Nabin Kumar (2020), 'Diaspora and Foreign Policy: A Global Perspective', *International Journal of Political Science (IJPS)*, 06 (04): 12–20.

Lakshmi, Rama (2015), Narendra Modi urges the Indian diaspora to become an extension of foreign policy, 2nd March 2015. accessed from The Guardian. http://www.theguardian.com/world/2015/mar/02/narendra-modi-india-overseas-diaspora-united-states

Lall M.C. (2001), India's Missed Opportunity—India's relations with Non-Resident Indians, Aldershot, Hampshire: Ashgate Publishing Ltd.

Lickinson, Jen (2015), Articulating an Indian diaspora in South Africa: The Consulate General of India, diaspora associations and practices of collaboration, *Geoforum,* 61: 79–89

Maini, Tridivesh Singh (2015), Indian Diaspora and the Restructuring of Diplomacy, Diplomatist.

Meer, F., 2003. Speech to the plenary session with the Ministry of External Affairs,9th January. In: Proceedings of the First Pravasi Bharatiya Divas, 9–11th January 2003. <http://indiandiaspora.nic.in/ppbd.htm.

Ministry of External Affairs (1990–1991), Annual Report, New Delhi.

Ministry of Commerce (2025), Ministry of Commerce: Export import data bank, accessed from https://tradestat.commerce.gov.in/eidb/iecnt.asp on 17th January 2025.

Ministry of External Affairs (2015), India - South Africa Relations, accessed from http://www.mea.gov.in/Portal/ForeignRelation/South_Africa_2015_08_28.pdf on 3rd January 2016.

Ministry of External Affairs (2024), India - South Africa Bilateral Relations, accessed from https://www.mea.gov.in/Portal/ForeignRelation/Bilateral-brief_SA-for-XPD-June-2024.pdf on 12th January 2025.

Ministry of External Affairs (2025a), Indian Missions Abroad, accessed from https://www.mea.gov.in/indian-missions-abroad-new.htm on 17th January 2025.

Ministry of External Affairs (2025b), Pravasi Bharatiya Divas, accessed from https://www.mea.gov.in/pravasi-bharatiya-divas.htm

Ministry of Overseas Indian Affairs (2015), Annual Report 2014–15, Government of India, accessef from https://www.mea.gov.in/images/pdf/annual-report-2014-15.pdf on 12th May 2020.

Ministry of Overseas Indian Affairs, (2009). Pravasi Samman Awards.<http://moia.gov.in/services.aspx?id1=30&id=m2&idp=30&mainid=23>(accessed on 27 December 2015).

Ministry of Overseas Indian Affairs, (2010a), Pravasi Bharatiya Samman Awards for14 NRIs. <http://overseasindian.in/2011/dec/news/20112212-111048.shtml accessed on 29 December 2015.

Ministry of Overseas Indian Affairs, (2010b),'India's Diaspora should Build Ties', Accessed from http://overseasindian.in/2010/oct/news/20100410-104658.shtml.

Modi, Renu (2010), Indian communities in Africa', Africa Insight, 40(1): 49–61

MOIA 2015, accessed from http://moia.gov.in/accessories.aspx?aid=10 on 28 December 2015.

Mukhopadhyay, Jayita (2010), 'Indian Diaspora in South East Asia: Predicaments and Prospects', *The Indian Journal of Political Science*, 71(3):995–1002.

Ndbele, S., (2004), Speech at a Cocktail Function Welcoming the President of India to KwaZulu-Natal, 16th September 2004, <http://www.polity.org.za/article/ndebelecocktail-function-welcoming-president-of-india-to-kzn-16092004-2004-09-16, (accessed 27 December 2015).

Parker, Darren (2023), Vedanta aims to complete $1bn investment in South Africa soon, Mining Weekly, 29th August, Accessed from http://miningweekly.com/article/vedanta-aims-to-complete-1bn-investment-in-south-africa-soon-2023-08-29 on 12th January 2025.

Preben Kaarsholm (2012), Diaspora or transnational citizens? Indian Ocean networks and changing multiculturalism in South Africa, Social Dynamics Vol. 38, No. 3, 454–466, p. 459

Radhakrishnan, S., (2005), "Time to Show Our True Colors": the gendered politics of "Indianness" in post-apartheid South Africa. Gend. Soc. 19, 262–281

Radhakrishnan, S., (2011), Appropriately Indian: Gender and Culture in a New Transnational Class. Duke University Press, Durham

Ramsamy, E., (2007), 'Between non-racialism and multiculturalism: Indian identity and nation building in South Africa', Tijdschr. Econ. Soc. Geogr. 98: 468–481.

Statistics South Africa, (2010), Mid-year Population Estimates. Accessed from www.statssa.gov.za/publications/P0302/P03022010.pdf on 2 January 2016.

Sudhi,K S(2024), 'Indian Diaspora makes it big in South African Parliament and provincial legislatures', 18th June 2024, The Hindu, accessed from https://www.thehindu.com/news/international/indian-diaspora-makes-it-big-in-south-african-parliament-and-provincial-legislatures/article68296657.ece on 14th January 2025.

The Economic Times, (2012), Indian SMEs, South Africa to work on enlarging small business base, 11th July 2012, accessed from https://economictimes.indiatimes.com/small-biz/sme-sector/indian-smes-south-africa-to-work-on-enlarging-small-business-base/articleshow/14818257.cms?from=mdr on 25th January 2025.

The Hindu, (2014). Madhya Pradesh CMPledges Rs. 1 Crore to Revive Mahatma's Tolstoy Farm. Accessed from http://www.thehindu.com/features/friday-review/history-andculture/madhya-pradesh-cm-pledges-rs-1-crore-to-revive-mahatmas-tolstoyfarm/article6101066.ece on 24th June 2019.

The Hindu, (2025), 'Indians abroad: On the 18th edition of the Pravasi Bharatiya Divas', accessed from https://www.thehindu.com/opinion/editorial/%E2%80%8Bindians-abroad-on-the-18th-edition-of-the-pravasi-bharatiya-divas/article69096100.ece on 27th January 2025.

The Times of India, Indian origin politician becomes South Africa's finance minister, 14 December 2015.

The Wire (2021), 'India Offers Vaccines, Ventilators and Research Collab to Africa to Combat 'Omicron', accessed from https://thewire.in/diplomacy/india-south-africa-omicron-help on 20th April 2024.

Tripathi, Ambikesh (2023), 'Growing Together with Perennial Bond: 30 Years of India-South Africa Relations', Organiser: Voice of the Nation, Accessed from https://organiser.org/2023/04/10/168354/bharat/growing-together-with-perennial-bond-30-years-of-india-south-africa-relations/ on 17th April 2023.

U. Mahajani (1976), "India and the People of Indian Origin Abroad", in M.S. Rajan, ed., *India's Foreign Relations during the Nehru Era*, Bombay: Asia Publishing House.

Vahed, G and A. Desai (2010), Identity and belonging in post-Apartheid South Africa: the case of Indian South Africa, *Joumal of Social Science*, 25, 1–12.

Malagasy Diaspora: Translucent but Palpable in Its Engagement

Rafalimanana Jennie

Introduction

In these first decades of the Millenium, International migration is always among world breaking news and it is actually within the globalization process. Madagascar, known as the big island in the Indian Ocean, has also recorded many of its citizens abroad and the Government is aware of the benefits of international migration for the country's economy. Therefore, it has adopted in its policy the concept of mainstreaming the Diaspora for its contribution to the country's socio-economic development.

Actually, Madagascar is at its starting point of harnessing the positive aspects of International migration and the Government has just decided to include its nationals living abroad in the country's development process. At the moment, this article does not make the difference between resident and non resident migrants yet. Notwithstanding their status, their ethnic, their work..., all of Madagascar's children abroad are gathered in the so-called Malagasy Diaspora. This perception comes from

R. Jennie (✉)
Embassy of Madagascar, Algiers, Algeria
e-mail: thinajuk@yahoo.fr

A. Dubey and S. Rajwar (eds.), *Diaspora in Development*,
https://doi.org/10.1007/978-981-96-6747-5_8

the fact that they spread throughout the world whose identity is unique and common, they all come from one country: Madagascar.

Hence, this paper will, firstly, relate the Malagasy citizens' immigration, then it will illustrate the Malagasy Diaspora's identification and profiling. After that, the paper will show the Diaspora's non-governmental engagement and activities, and finally, it will project cooperation between the Malagasy administration and its Diaspora.

Story of the Malagasy Citizens' Immigration

Each migrant has his reason to leave his country: for studies, works, family reunification, slavery, conflicts, and so on. Madagascar has its migrants all over the world and 169 984 people outline the Malagasy Diaspora in the five continents (United Nations Department of Economic and Social Affairs 2015). It is worth noting that this figure does not include Malagasy migrants residing in some countries like the United States of America, Lebanon, the United Arab Emirates (UAE), Saudi Arabia, Kuwait (these countries actually welcome too many Malagasy workers at the present time) and African countries such as Djibouti, Kenya, DR Congo, Gabon, Cote d'Ivoire, Senegal, and so on.

Malagasy diaspora has been built in several flows of more an emigration of knowledge than an emigration of work (Pitchboule 2009). Unlike migrants from Africa and the Maghreb who originally exported in group to meet the labour needs of industries and services in Northern developed countries, Malagasy migrants moved to Europe in order to get higher education because Malagasy society has always considered knowledge as an essential value.

Actually, this determination to acquire knowledge abroad for a social recognition is attached to the perception of former Malagasy immigration in the imperial period. The first envoy began during the King Radama I era, in nineteenth Century while some Malagasy people were sent to England for studies. At that time, teaching and schooling were of a certain value in the following Kings' era and too many schools and churches were opened throughout the country. Some migrants travelled also to France for studies which had stopped in 1916 because of the Nationalist movement called '*Vy Vato Sakelika*'. Besides, it is said that 40 000 hairy men had been enrolled for the First World War which composed the first migratory wave. Others went there for a labour migration and as skilled workers in armament or aeronautical firms (Dornel 1995). Malagasy migrants at that time were almost sent abroad for military trainings.

Then, many officers and riflemen were sent to France again during the Second World War and have got the right to be French people. Most of them stayed in France whereas some returned back home. Some of Malagasy students in Medicine and young folkloric artists took part in the French colonial show in 1931 and stayed there. Most of authorities' children could also go to Europe for medical studies and theology. Apart from that, nationalist students who were of noble family descendants were to be exiled in France by colonial administration during 1947s event and were spread throughout France to avoid nationalist group movement. However, some Malagasy students gathered and created an association for students of Madagascar origin (AEOM) to talk about assimilation to France (Rabenoro 1975).

Later, around Madagascar's independence, in 1960, always from the thirst of knowledge, some students of high school went to France for higher education so that they could get French diploma and became at the same level as French cadres (Crenn 1998). The main goal was also to obtain high position when coming back to home country. The period of 1945 to 1975 is called '*Les Trente Glorieuses de la migration*' or 'The glorious thirty of migration' (Blanc-Chaleard 2001).

In the 1970s, notwithstanding numerous returns back home, many Malagasy citizens left the country because of political and social problems. This was emphasized by the change of education system which has taken the Malagasy language as the education language. Known as '*Malgachization*', it was considered as a claim for identity but was actually perceived by young people as a deadlock which provoked them to emigrate. These Malagasy migrants between 1975 and 1990 were designated as the second generation of migrants by Crenn. They emigrated and once arrived in France, they decided to stay there.

Thus, emigration of Malagasy citizens abroad has got too many reasons and continued in the 1990s. This last generation of migrants composed of 'students-workers' (Crenn 1998) were much attached to the culture. Emigration to Western region was a Labour migration. These migrants went on taking some professional training in order to easily integrate the Labour market. Family reunion came also to thicken their number which reason was likewise to satisfy Family's ambitions for material, intellectual and social success.

Malagasy emigration increased in the second decade of the 2000s because of politico-economic crisis. At the present time, the brain gain which is the reverse of the brain drain is favourable to support the development process. The increasing number of the Malagasy Diaspora has relatively reached its capacity on which Madagascar can lean on and their competency would be a capital privilege to implement the General Program of the State.

France as it has a historical link with Madagascar is the most chosen destination of Malagasy migrants (Metropolitan, DOM-TOM and especially in La Réunion). The approximate figures give about 25 000 to 150 000 Malagasy citizens in France (Pitchboule 2009). Later, they fly to Switzerland, Germany, Belgium and Norway as Europe was considered formerly by Malagasy migrants as idyllic. However, it has not lasted long because of many issues relative to the stay (visa, duration, work …). Then, the wind of American dream blew in Madagascar as the United States is a symbol of success, migrants opted to go there, in Canada or in Africa as well.

A European anthropologist said that there are neither statistical studies nor analysis about Malagasy diaspora. Therefore, Malagasy migrants tend to be a '*translucent Diaspora*' before host countries. However, the new Directorate within the Malagasy Ministry of Foreign Affairs under the supervision of the Directorate General of the economic promotion and the diaspora was created to frame from now on the Malagasy diaspora profiling.

It is to be noted that there are available data on Malagasy Diaspora in the Ministry of Foreign Affairs though the figures from the immigration service of the host countries and that in the Madagascar Consular services are different. This is because, only few members of Malagasy diaspora come to register in the Consular services. For instance, in Ethiopia, 45 Malagasy people are recorded at the immigration service while only 16 have been registered at the Malagasy embassy in Addis-Ababa. It is like in La Reunion where 5000 people are registered at the immigration service and only 1291 are enrolled in the Consular service. And in France, there are approximately 107 000 Malagasy migrants listed at the immigration service but only 1618 of them are matriculated in the Embassy. In total, there are about 121 398 Malagasy migrants recorded in these host countries and only 9091 are actually inscribed in the Embassy's database.[1] This really leads researchers and observers to qualify the Malagasy Diaspora as *translucent* for both of the host and origin countries.

To adjust the above discrepancy in the Malagasy Diaspora numbers and to adopt the globalization concept of harnessing Diaspora's potential, it is worth approaching the Diaspora. Thus, throughout its roadmap, the Directorate of Diaspora plans to identify the Diaspora in order to make its profile and to get data about them.

Malagasy Diaspora Identification and Profiling

From different researches, while the Malagasy diaspora was born from various ways and in different periods, there has not been any way to characterize the Malagasy Diaspora yet. However, they remain Malagasy citizens who have their country, their land, their origin somewhere and have their identity. Nowadays, since the higher education in the country has been improved, searching for better life by working abroad is the main objective of all migrants, Malagasy nationals included.

It is obvious that most of them are skilled or semi-skilled migrants. This promotes the 'brain drain' but this potentiality of human resources abroad can be strength to the home country's economic development. In fact, Malagasy migrants have got a good shape of being 'best foreigners' because of their know-how, adaptability and culture which make them at ease and safe in European host countries. It protected them from classic stigmatization (Pitchboule 2009).

Madagascar, an island of 23.000.000 inhabitants in the Indian Ocean, a developing country, has officially recorded in its external representations only about 9091 of its citizens abroad, at the moment. Whereas the UNDESA, in 2015, found 169 984 Malagasy migrants in the world, IOM after FORIM's survey, in 2016, estimates that Malagasy Diasporic potential between 100 000 and 140 000 people. Nevertheless, projection tends to count about at least 10.000 Malagasy nationals—registered ones—spread all through the world. Most of them are students with a status of from six months to a long period of stay about three, four or five years. Some are migrant workers wishing to get back home sooner or

[1] SEBDS/DDIA (Ministry of Foreign Affairs, Madagascar), 2016.

later. Others are surely former students or workers having decided to be located overseas after feeling better life there. All of them are members of Madagascar's Diaspora within the International migration system.

As it is cited above, not half of the Malagasy diaspora have been inventoried in the database of the Ministry of Foreign Affairs. In fact, Diaspora can be known from more bad events occurred on them than good ones. These situations have particularly drawn Malagasy authorities' attention. Accordingly, Malagasy Diaspora needs much care, accurate assistance and adequate protection. Meanwhile, the President of the Republic of Madagascar, Hery RAJAONARIMAMPIANINA, has met some members of his compatriots, in marge of some International meetings. During each reunion, the Diaspora spoke about their life conditions and also declared their wishes for the country, whereas the President has shared to them about the country's expectations from all citizens.

Since then, the President has decided to incite them in taking part in the country socio-economic development by giving them this role. For so doing, conscious of the international migration's benefits for the country's economy, the Government has adopted the mainstreaming concept of the Diaspora in its policy for the Diaspora contribution to socio-economic development. This is in the framework of the new perception for the International migration and is among its positive aspects, not only in Europe but all over the world. Formerly, it has been denounced as brain drain, while nowadays, it is considered as an opportunity to be harnessed and is the way to reverse the concept of brain drain into *brain gain*. This new concept will give triple benefit. Firstly, for the migrants themselves as they can improve their life conditions; secondly, for the country of origin which can profit from its nationals going abroad, for example, the decrease of unemployment figures in one hand, and from their contribution in financial and social remittances or in investments in another hand, this will put the Diaspora as lever for the home country's development; and finally, the Diaspora, a group of foreigners will strengthen the host countries' manpower. The host countries will profit the migrants' expertise which fosters the country's development (Gubert and Giordano 2006).

Governmental Structure

From this new perception of International migration for the country's economic growth, and to actually acknowledge Malagasy Diaspora's weight in the home country life, the Malagasy government has established the new structure within the Ministry of Foreign Affairs in 2015. This unit is in charge of the Malagasy Diaspora known as the Directorate for the Diaspora under the supervision of the Directorate General for Economic Promotion and Diaspora. The Directorate for the Diaspora's duties are:

- To offer protection to all Malagasy citizens' interests and to provide support to those in difficulties. These are among the essential responsibilities of the Diplomatic missions which are part of the Ministry of Foreign Affairs.
- To undertake best coordination and efficiency of the Government's actions: establishment and implementation of the National policy for the Diaspora's engagement and setting all suitable measurements for the Diaspora's mobilization in its participation in the socio-economic development of Madagascar, along with other Departments.

There are four services in this Directorate:

1.Service for censing and profiling the Malagasy Diaspora
2.Service for the assistance and protection of Malagasy nationals abroad,
3.Service for the partnership, mobility and migrants workers.
4.Service for mobilization and return facilitation for full participation of Diaspora to the sustainable development of the country.

This new organization aims to improve assistance and connection between the Diaspora and the Government and to better coordinate stakeholders' actions. As it is still at an embryonic phase, diaspora mapping, trust rebuilding and the development of a national policy for diaspora's commitment are among its short-term priority. This Diaspora's commitment will set a better environment and a long-term strategy in favour of investment, financial and social remittances from the diaspora.

Roadmap of the Directorate for the Diaspora

In order to perform these activities, the Directorate has launched its roadmap which shows five big parts as the following: (1) Capacity building; (2) Better understanding of the diaspora; (3) Reinforcement of the link between the diaspora and Madagascar; (4) Adoption of measures to promote diaspora involvement in the socio-economic development of the country (5) Consolidation of cooperation.

Capacity Building

The Directorate in charge of the Malagasy Diaspora is the first governmental entity established for the first time in Madagascar. Thus, considering all the issues and challenges relative to the Diaspora's life, this first part is essential to bring all the direct actors of international migration and diaspora management system on track. These actors are Directorate staff, consular officials in the external representations and members of the interministerial committee in charge of the establishment of the diaspora's engagement.

On July 2015, one-day workshop entitled: 'Mobilization of Diaspora' was held in the Ministry of Foreign Affairs which was organized jointly by the Ministry and the International Organization for Migration (IOM 2016). It was actually about migration and development with knowledge of best practices for diaspora's engagement.

To better understand these practices, members of the Directorate staff have participated in any course relative to this topic. Thus, on September 2015, the Chief Service for Malagasy citizens abroad attended an International Migration Law course in Sanremo, Italy. This course is particularly based on legal instruments to undertake for migration management and particularly for the protection of migrants.

Meanwhile, a free course of six weeks was available online in October and November 2015 in which Directorate staff have also participated. This was about international migration and development.

In addition, Personnel members of this Directorate have attended the world forum on migration and development in Istanbul, Turkey in October 2015 and in Dhaka, Bangladesh in December 2016.

All events about migration and development or Diaspora's engagement towards its home country would be very interesting for experiences sharing and new knowledge acquirement. Thus, the Indian Organization for Diaspora Initiative (ODI) holds every year an International conference about International migration and the Diaspora's engagement. This organization works to understand the status and role of Diasporic communities across the globe. Such conference can actually help Malagasy structure for Diaspora advancing for knowledge acquirement and experience sharing. Thus, in January 2017, the Chief Service for the Malagasy citizens abroad has fully participated in this ODI's international meeting in Bengaluru, India.

Better Understanding of the Diaspora

Notwithstanding the will to harness the Diaspora's potentiality for the country's development and to assure the policy's achievement, it is crucial to get a particular approach to the Diaspora as to know them well and to consider their link to the country of origin.

The first and the biggest assignment in this part was the census of all the Malagasy Diaspora in order to know their number, locality, activities, point of view about their role (money transfer, return...) etc.

So, the Ministry of Foreign Affairs has initiated a project for the study of the Malagasy Diaspora's profiling in collaboration with the IOM and the Embassy of the French Republic in Antananarivo Madagascar. The study has been conducted by FORIM from May to July 2016. The outcome of this study has been delivered during a restitution workshop held at UNESCO Paris, France, on 01st September 2016 in which the Malagasy Ministry of Foreign Affairs, the French Embassy in Antananarivo, and some members of the Malagasy diaspora and associations have participated.

Since the Directorate for the Diaspora is a new system in the Ministry, this latter has chosen France as the experimental country because of the increasing numbers of the Malagasy diaspora residing there and, for the fact that there is an Embassy of the Republic of Madagascar in Paris. The survey has revealed a huge number of about 140 000 Malagasy migrants. *Questionnaires* have also been displayed online by the Central Department of Foreign Affairs in order to identify the Diaspora's opinions and expectations relative to the country's interests. In the meantime, consular services have improved their delivery in order to incite the Diaspora to, first of all, register and to be used to get in touch with the Embassy.

This has allowed the Directorate to draw the Diaspora's profile and map, and to keep it updated. Later on, study of Diaspora's profile in other countries will be taken in charge and the next country will be Canada.

Reinforcement of the Link Between the Diaspora and Madagascar

This part is about the re-establishment of Diaspora's trust towards the state, familiarization of the Directorate with them and sensitization on their role in the country's development.

Diaspora's commitment to its country of origin mainly depends on the link between them. In addition to the Consular services' distribution, it is very necessary to promote and support all activities for Malagasy culture and identity such as 'Madagascar's Day at UNESCO'. Both the Diplomatic mission and the Diaspora will work together for this promotion.

For so doing, the Diaspora needs to be sensitized about the country's requirements and opportunities. Therefore, the Directorate during its mission in France, in September 2016, has led diverse meetings in France with different French entities working for co-development or migration and development. Malagasy Embassy and Consulate's Officers have taken part in these meetings and particularly with the Malagasy Diaspora in France. The last meeting was about introducing the recently created Directorate's structure and activities. It also allowed all the participants to share ideas (Diaspora's requests: right to vote, consular registration, investments to Madagascar and customs conditions, etc.), and to discuss about future cooperation.

Anyway, it should be always reminded that the Directorate has just been created in 2015 and from its roadmap, some activities are still in their primary steps. However, some projects have already been implemented whereas others will be established.

As far as the improvement of the consular services' delivery is concerned, online services are available for public which are very useful for all the Malagasy Diaspora living far from the Embassy. A hypertext link through the Ministry of Foreign Affairs' website has also been set up to make useful data accessible, and to share information in time. Moreover, regional offices for Malagasy passport delivery within some Embassies, in the United States, in Africa, in Europe and in Asia, are going to be opened. It will avoid coming back home to renew the passport or waiting for long period of the new passport's expedition. And finally, in order to assure better assistance and protection of Malagasy citizens abroad, it is fundamental to appoint a responsible for them within each external representation of Madagascar, who will work closely with the Directorate for Diaspora in the Central Department.

Besides, in the attempt to promote the Diaspora's rooting into Malagasy identity and culture, especially for young generation, the creation of '*Madagascar House*' in more than one pilot country is among the projects. Cultural activities such as shows, education (Malagasy language, dance, speech, traditional games ...) or contests will be organized and supported in this framework where cooperation with Diasporas' associations would be privileged and improved. Further, a volunteer program for youth will be launched for experience sharing and promotion of Madagascar.

A consular seminar would have taken place but Madagascar has welcomed more than one Summit in 2016, of COMESA (Common Market for Eastern and Southern Africa) and of FRANCOPHONIE (French-speaking countries' Organization). The Ministry of Foreign Affairs has got full of activities for the preparation of those big events. In 2017, the Directorate will hold a Diaspora Forum in the Capital of Madagascar during which all Malagasy Diaspora are invited to participate.

Two other events have to be profited as well which are the Madagascar's day in UNESCO and the '*Rencontre Nationale Sportive*' (RNS) or National Sportive Reunion, organized by the Malagasy diaspora in Europe.

Adoption of Measures to Promote Diaspora Involvement in the Country's Socio-Economic Development

Development of national policy for Diaspora engagement is the greatest challenge for the Directorate because all actions towards the Diaspora are based on it. It requires the creation of an inter-ministerial committee which not only will draw up the said policy but will legally study the means of its implementation such as the various incentives to be adopted. It also depends on the Diaspora identification and profiling which have begun next to the Malagasy Diaspora in France. Next profiling will be in favour of the Malagasy Diaspora in Canada. It is to be underlined that the task remains grandiose.

Consolidation of Cooperation

Given the complexity of problems and difficulties encountered by Malagasy migrant workers abroad recently, it is necessary to establish a framework for cooperation with various host countries in order to improve measures to protect those migrant workers. And this is among the Malagasy Government's objectives in this consolidation of international cooperation. Since the Government makes an effort to approach these Middle East countries in order to actually set up agreement, Madagascar is at the present time in negotiation with Saudi Arabia and Lebanon for the provisions within such bilateral agreement.

For multilateral cooperation, Madagascar's membership to international organizations working within the migrant protection framework should be made profitable. It may allow Madagascar acquiring project funds and capacity building such as with the International Labour Organization (ILO) and the International Organization for Migration (IOM 2016).

Hence, with this Directorate of Diaspora's roadmap, the Malagasy Government is endeavouring to improve its support to the Diaspora by:

- Censing and profiling the Malagasy Diaspora and getting its data
- Sensitizing the Malagasy Diaspora to approach the Embassy or Consulate in order to register and to tie relationship with the country's representations
- Computerizing of consular service
- Holding an '*Action for Passport*' for its renewal within Madagascar embassies or regional office
- Calling for a consular conference and a Diaspora forum
- Completing the National policy for the Diaspora's engagement within the inter-ministerial committee
- Mobilizing the Diaspora and supporting its associations' activities
- Ensure bilateral cooperation with countries where Malagasy compatriots live and work
- Searching for cooperation with private sectors for any projects to be sponsored by the Diaspora

Diaspora's Non-Governmental Engagement and Activities

As already stated previously, the structure for taking charge of Diasporas' matters is very new. However, this is far from considering the Malagasy Diaspora as young and passive. The Malagasy Diaspora and its associations have existed long ago, so as their activities which have been led outside the governmental framework. They have done much on their own long before the new structure of the Government has been created.

Malagasy Diaspora's effort for gathering began in 2000 with the '*Hetsika Diaspora*' (Diaspora action) through a virtual network composed of Malagasy nationals in North America and France (EGDM 2015).

Since its existence, the Malagasy Diaspora has got its identity and activities. It acts for its country of origin (Madagascar) through its associative life which intervenes in social or humanitarian fields through family ties and investments. Diaspora actually cares about its families in the home country and has already had their part of stone to build the Nation in order to participate in its development.

Meanwhile, without any accurate figures for the Malagasy Diaspora's numbers, the Malagasy Diaspora is said to be a virtual community (Pitchboule 2009). Nevertheless, it has a great potentiality for the country's development. Its members, for their proficiency, their political culture and their professional success can take big parts in such contribution. This potentiality only needs to be mobilized, organized and well developed beyond the initiatives and actions already carried out. Moreover, it is clear that the new generation truly aspires to '*a return to the soil of their origins*' and it is a real claim for identity of these young Malagasy people who live outside their country of origin. It has been reported, in 2009, that these young people actually want to '*act*' instead of '*just appear*'. Thus, six years later, the Government would like to take in hand their aspirations and to support their activities by setting the governmental structure in charge of them within the Ministry of Foreign Affairs of Madagascar.

In fact, Malagasy Diaspora on behalf of its associations and/or individuals have already worked for socio-economic cares towards their countrymen in different fields in a non-governmental way. In this article, some Diaspora associations, groups or events have been selected for illustrating Diaspora's actions for its compatriots, such as *Amboara*, *Zama* (*ZAnaka Malagasy Am-pielezana*) within the *Etats Généraux de la Diaspora Malgache* or General State of Malagasy Diaspora (EGDM) which initiatives are in favour of Diaspora's members, supporters and associates..., GasyZah, GasikArt, *Environnement Développement Action* (ENDA), Malagasy Club de France, Religious associations and *Fanjava*...

Amboara: it is a cultural, sportive and humanitarian association based in France. It works for schooling children in Madagascar with 22 volunteers. They perceive education as a compulsory way to attain society development and for the whole country. They have targeted Madagascar for their roots and because they have ascertained that education level in the country is still weak as compared to which in other countries. Among their activities, there are renovation of schools, donation for school kits, materials… They have also given financial support to other associations such as the '*Masoandromody*' which works for sensitizing people about Alzheimer and looks after old people affected by this disease.

The *EGDM initiative* has been launched in 2015 by two of '*Hetsika Diaspora*' founders (from France and Canada). It is now composed of Diaspora members from Canada, the United States, South Africa, France, Italy, Sweden, Germany, Dubai, Poland, Seychelles, La Réunion and DR Congo. Their motto is '*Hery mitambatra no zotra filamatra*' or join strength as the force idea.

From the '*Hetsika Diaspora*' on, articles in daily magazine and Radio or TV programs began to focus on Diaspora's matters which are available online. It has been followed by the creation of Facebook page called '*Preparation des EGDM de la Diaspora de Madagascar*' for easy communication in order to prepare the *EGDM. Questionnaire* towards Malagasy Diaspora has been launched. Diaspora Website of each country where the Malagasy diaspora are based has also been established which is in the National language.[2]

Later, in May 2015, the '*meeting of Malakoff*' took place during which some Diaspora members have joined through Skype. Then, organization committee has been created to prepare the EGDM (2016) which has been held in Aix-en-Provence in July 2016. The rehearsal took place in Wakefield Canada on 13–14 May 2016.

In fact, the *EGDM* initiative is conscious of the home country's needs and favourable to Diaspora's contribution. Its members have adopted its name as *Zama* which stands for *ZAnak'i Madagasikara Am-pielezana* and which means Madagascar's children abroad, or Diaspora of Madagascar origin instead of *Malagasy Ampielezana* translated as Malagasy Diaspora. This choice is likely to justify their will to work for Madagascar with inclusiveness. In other words, they want to include everyone from Madagascar: Indo-Pakistanis Malagasy, Chinese Malagasy, French Malagasy, '*Zanatany*' (foreigners born in Madagascar), and foreign spouses of Malagasy citizens to contribute to the country's development. Their vision tends to be '*a solidarity-based Malagasy Diaspora by its exemplarity*

[2] http://malagasy-canada.org.

and actions, makes itself more useful to itself and to its country of origin' (EGDM 2015).

Consequently, *Zama Aix 2016* has been held in Aix-en-Provence, France, from 8 to 10 July 2016 which has been an attempt to assemble Malagasy Diaspora of the world. Among others, a young Malagasy of 23 years old has presented her project entitled '*L'arbre qui pousse à l'emploi*' (The tree that grows for employment) affirming that the development of Madagascar can also pass through a million of trees planting by 2020. She is at the head of a company known as '*Alamanga Reforestation*' with several dozen seasonal workers. Her company aims at preserving biodiversity, combatting climate change and promoting changes through gentleness with efficiency and firmness of Madagascar population. She is confident that Malagasy people have more opportunities than they could not imagine, they have to act and harness all what they have in their possession. She launched: '*Comment quand on a tout peut-on n'avoir rien?*' (How can we get nothing while we get everything?). This event in Aix-en-Provence is a preliminary of the EGDM project of 2019.

Another event that gathers all Malagasy diaspora is the well-known *RNS which stands* for '*Rencontre Nationale Sportive*' (National sports reunion). In 2016, it was held in Lyon, France. Initially, RNS was only for sports as it is so-called, but it becomes a socio-cultural assembly later. It actually allows Madagascar people abroad meeting in one location. Thus, cultural activities began to take place for everyone's happiness.

Activities held by Malagasy cultural associations like HAVATSA-UPEM (Union of Malagasy poets and writers) or FIMPIMA (Malagasy orators' associations) and AIFM (International association of women from France and Madagascar) reflect the Malagasy cultural identity which revives the Malagasy Diaspora's feelings towards their home country within culture and language.

The President of a Diaspora association based in Switzerland reported that cultural events dominate the Malagasy Diaspora activities in Europe in recent months (2016). This has been reflected by the arrival of the Malagasy group singer AMBONDRONA in Switzerland for two weeks, Joseph DAF in France and many other artists as well who create pleasant atmosphere for those Malagasy migrants known as '*mila ravin'ahitra*' (searching for good). This situation, as always, gathers the '*gasy d'andafy*' (Malagasy abroad/overseas) and generates a bizarre feeling of belonging to something: '*Tanindrazana*' (Motherland) where grandparents still live for some, 'dad and mom' for others. In short, artists come there to remind them that they are not alone in Europe but are part of a community known as the Malagasy Diaspora. This is one way to revive their connection to the home country. During such cultural show, *Gasy'zah*, a Diaspora's association based in Valleiry France, held other actions such as selling donuts in order to collect funds for the association's activities.

Also, Radio '*Gasikart.com*' from Toulouse came to Paris to cover the *Conference des Bailleurs* (CBI) or *Backers Conference* organized by the Presidency of the Republic of Madagascar in Paris (1–2 December 2016). Thus, despite the cold winter, the Diaspora in Europe is very active lately.

Still in this cultural circle, the *Malagasy Club France (MCF)* is a Madagascar corner for the Malagasy Diaspora in France where many clues to discover, such as Malagasy cuisine, language, customs etc.

As far as religious matter is concerned, there is, for example, the alliance for Malagasy protestant churches in France *(FPMA: Fiangonana Protestanta Malagasy aty Andafy)* and the Malagasy national catholic almonry in France (*ANCMF*: *Aumonerie Nationale Catholique Malagasy* de France). They actually tend to unify Malagasy group with common spiritual identity and for community activities.

ENDA Tiers Monde is an international NGO founded in 1972 and based in Dakar, Senegal. Since the 1980s, it has been carrying out specific actions for development in Madagascar. In August 1996, ENDA Indian Ocean signed a headquarters agreement with the Malagasy government and launched three development programs with the support of the various partners and the collaboration of other NGOs already present in the field. It has launched the DIAPODE program or Diaspora for the Development. It conducts a study about the Diaspora, their initiatives and partners.

All these cases show that the Diaspora and its associations endeavour to reinforce the link between its members for the common values and experiences which they want to transmit to the next generations.

Last example but not the least among Diaspora's associations' activities is about a platform known as *Fanjava* which is led by two brothers for crowdfunding. They endeavour to give support to the Malagasy young generation. Their objective is to incite and allow Malagasy diaspora and Madagascar friends to finance Malagasy startuppers' project through their platforms like Gofundme, Kisskissbankbank and Ulule. *Fanjava* was present at *Zama-Aix 2016* and the platform Ulule permitted to finance *Niouz* Madagascar project for its participation to Africa innovation contest. As a result, *Niouz* Madagascar has collected 600 Euros to carry on its exploit. *Fanjava* platform wants to go on creating a network for competency, accompaniment, placement and refinancing Initiatives with Malagasy investment funds (Koolsaina 2016). Their slogan is '*Doing more and better with less from Malagasy, for Malagasy*'.

In the meantime, apart from these socio-cultural activities of Diaspora and its associations, their economic contribution is rather non negligible. Human migration has actually an impact on countries' economy because migrants' financial remittances would be a big size (Gubert and Giordano 2006). Besides, the 1990s and the beginning of the Millennium confirmed the increasing role played by the migration in context of the economy globalization (Garson and Mouhoud 2017).

The World Bank (2014) stated that the Malagasy Diaspora made remittances of USD 547 million to Madagascar in 2010 and USD 427 million in 2014 which are the equivalence of 4% of Madagascar GDP in that year.

From the IOM report in 2016, the Malagasy Diaspora in France make remittances to Madagascar estimated at Euros 86 million by individual and through associative initiatives in collective in 2013. These financial transfers were carried out by 70% of the respondents to the Questionnaires during the survey which revealed an average amount of 155 Euros per month and which mainly serves to support the family budget in Madagascar. Besides, there are more than 240 associations of Malagasy nationals created during the last ten years. It seems that half of the members of the associations are directing their involvement in solidarity initiatives in Madagascar, particularly in the education and health sectors.

Though the Malagasy Diaspora is seen as translucent for the fact that they constitute a big network and bring huge parts to the country's development, their activities need to be more mobilized and framed. Thus, it is better to get innovative approach to the Diaspora and to establish a real cooperation for the common Homeland's bright future.

Towards an Effective Cooperation Between the Government of Madagascar and Its Diaspora

Malagasy Diaspora in the world and all Madagascar friends can institute economic and financial strength for Madagascar. All of them recognize their potential and are convinced to be an important competency fishpond and lever of Madagascar's socio-economic development. Most of the Malagasy Diaspora members are determined to do much for the country but their activities need to be mobilized and developed. Accordingly, it is opportune to connect the Diaspora's wish to contribute to the country's development with the country's expectations, and the Government's will to give them this role.

In fact, it is to be emphasized that the Malagasy Diaspora have done their best but on behalf of the common Motherland, all Malagasy people can collaborate for the common goal as the country's socio-economic development. Therefore, as it has been already developed in the previous chapter, social and financial remittances have already been existed. All of figures about remittances actually give hope to the country's economy if there is an accurate framework of cooperation between the Malagasy Diaspora and the Malagasy Government.

Among the Directorate for Diaspora's missions, in collaboration with other involved technical departments and in consultation with the Diaspora itself, the elaboration and implementation of the National policy for the Diaspora's engagement remains the big step to make this cooperation completed.

It is always heard that Malagasy diaspora is reluctant to work with the Government and denies its tutelage. They can still go on developing their actions without it. The committee for *EGDM initiative* confirmed that *EGDM initiative* is 100% apolitical, secular, civic/citizen, independent while being mindful of Malagasy values, etc. (EGDM 2016). And even the *Zama-Aix* event has underlined its wish to be independent from the Government or from any other kind of tutelage (Rabetafika 2016). However, it is worth noting that the actual role of the governmental structure is neither to supervise nor to lead the Malagasy Diaspora but to motivate and support them in their activities. The objective is that everything can be done in a formal way such as the money transfer or goods expedition.

After all, the Government has responsibilities towards its nationals and Madagascar's diplomatic/consular representations in a given country will perform the protection of the accrediting State's interests and of Malagasy nationals residing in the hosting State as it is stipulated in the Vienna Convention for Diplomatic Relations Art.3 Al.1.b. This cannot be changed and Consular services which delivery has been just improved are there, for assistance and protection of all Malagasy migrants. Furthermore, the Directorate for Diaspora in the Central Department continues to honour its duties which are to take in charge of Malagasy citizens abroad: the Diaspora.

Diaspora engagement policy is then the legal framework for this cooperation. The coordination of the Government program with the Diaspora's will to act is a must to set a climate of trust between the Government and the Diaspora. Again, it should be stressed that the mobilization of and the support to the diaspora are at its early stage but will continue as this cooperation progresses. Besides, the first liability of the Diaspora is to be registered in the representations of Madagascar so that they can be identified and be provided on time assistance and support, where applicable.

Conclusion

The Government has already recognized the Diaspora's strength and potentiality on which the country can depend. Together, the Government and the Malagasy Diaspora can undertake a productive cooperation for the country's development. And the new orientation of the country's foreign policy tends to lean on the Diaspora's leading role. Thus, the national policy for Diaspora's engagement will target Diaspora's skills and talents which need to be mobilized and harnessed. The more Malagasy Diaspora are supported and their interests protected, the more motivated in all their activities for the country's development they will be.

References

BLANC-CHALEARD. 2001. 'Les Trente Glorieuses de la Migration'.

CRENN, Chantal. 1998. "L'espace migratoire franco-malgache. D'une migration temporaire à une migration définitive".

DIRECTORATE FOR THE DIASPORA, Directorate General for Economic Promotion and the Diaspora, Ministry of Foreign Affairs, Madagascar. 2015. "Feuille de route de la Direction de la Diaspora".

DORNEL Laurent. 1995. "Les usages du racialisme. Le cas de la main-d'œuvre coloniale en France pendant la Première Guerre mondiale".

EGDM. 2015. "Charte des valeurs EGDM". Septembre 2015.

EGDM. 2016. "Chronologie de la Démarche des Etats Généraux de la Diaspora Malgache (EGDM)". Canada.

GARSON, Jean-Pierre and MOUHOUD, El Mouhoub. 2017. "Migrations: Tendances migratoires actuelles".

GUBERT, F. and T. GIORDANO. 2006. "Migrations internationales et développement: un tournant? ". Synthèse, n° 02, 2006.

IOM. 2016. "Etude du profil de la Diaspora Malagasy en France". Report.

KOOLSAINA. 2016. "Fanjava: 1ere platfome de crowdfunding 100% Malgache". www.koolsaina.com.

MALAGASY CLUB DE FRANCE, 2015. Interview to Mr. Jean Razafindambo: "Des Etats Généraux de la Diaspora Malgache".

PITCHBOULE, Lalatiana. 2009. "La diaspora malgache a-t-elle une identité…?" Madagascar Tribune, 2 September 2009.

RABENORO, Auber. 1975. "Trois générations de Malgaches en France: trois formes d'adaptation? (1930–1945; 1945–1960; 1960…)".

RABETAFIKA, Roger. 2016. Réforme fiscale et révolution socialiste à Madagascar. Éditions L'Harmattan, collection Repères pour Madagascar et l'Océan Indien.

RAHAJARIZAFY, Lanto and RASAMIMANANA, Tivo 2016. "Restitution de l'étude sur le profilage de la Diaspora en France". Antananarivo / Paris: Ministère des Affaires Étrangères de Madagascar & Organisation Internationale pour les Migrations (OIM). Rapport final publié en octobre 2016.

RASOAMANANA, Vola. 2014. "Amboara Association". Madaplus

RATSIMBA Landry. 2016. "News from Malagasy Diaspora in Europe"

United Nations Department of Economic and Social Affairs. 2015. International Migration Flows to and from Selected Countries: The 2015 Revision. United Nations. Retrieved from https://www.un.org/development/desa/pd/sites/www.u

WORLD BANK. 2014. http://www.worldbank.org/en/topic/migrationremittancesdiasporaissues/brief/migration-remittances-data. Report.

The Private Sector, Diaspora Engagement and the Migration-Development Nexus: Comparative Cases from the Philippines

Satwinder Rehal

Introduction: A Long History of Movement by Filipinos

Garchitorena (2007) states that conceptually, "Diaspora" refers to the "historic mass dispersions of people with common roots including as well the 'transnational community' those people with a shared identity extending or going beyond the national boundaries of their country". When used in the Philippine context, the term is ineluctably global, characterizing the many countries where Filipinos are located and also in discursive discourses (Aguilar 2015; Ramos 2013). Garchitorena (2007) outlines key three distinctive historical movements of Filipinos; the first wave traced to the indentured labour migration of sugar workers from the Ilocos Region in Northern Philippines to Hawaii to work for the Hawaii Sugar Platers Association from 1906 through to 1929. Cruz (1994) notes that following a union strike in 1924 and subsequent ban on Filipino labourer in Hawaii, a shift of Filipino migrant labourers to the

S. Rehal (✉)
HZB School of International Relations and Diplomacy, The Philippine Women's University, Manila, Philippines
e-mail: ssrehal@pwu.edu.ph; dickush.rehal@gmail.com

A. Dubey and S. Rajwar (eds.), *Diaspora in Development*,
https://doi.org/10.1007/978-981-96-6747-5_9

US mainland occurred mostly in terms of farm workers to California. As a consequence of the Great Depression of 1929, Filipino labour migration to the United States slowed, precipitating seeking opportunities in other parts of the world, notably in the Netherlands were most were hired as seamen (Sicam 2003).

While the first wave of Philippine labour migration to the United States was essentially skills based recruitment in farms, factories and in military facilities, the second wave, occurring in the 1960s was characterized predominantly by professional cadre comprising doctors, nurses and engineers who migrated to support the booming economy while the US was involved in the Vietnam War (Garchitorena 2007). This wave of Filipino migration to the US was further enhanced by academic scholarships and exchange programmes which prompted many to permanently settle in the country. Garchitorena (2007) notes that the second save, unlike the first one, was smaller in demographic size and short-lived though it attributed to the start of "brain drain" phenomenon as many of the country's best and brightest graduates opted to settle away from the Philippines.

The third and most significant wave occurred in the early 1970s under the Marcos regime when the Philippine government began to actively encourage migration embodied in the 1974 Presidential Decree No. 442 of the Labour Code to mitigate against a high prevalence of domestic unemployment of an fiscal policy that sought to stabilize the country's dollar reserves. This labour migration policy witnessed the creation of certain regulatory government agencies, notably the Overseas Employment Development Board, and the National Seamen Board (Garchitorena 2007), which also oversaw Filipino labour migration to the Middle East in response to the need for skilled professionals to the Gulf to work in the surging petrochemical and infrastructure industries during the "Oil Boom" era spurred by rising oil prices. Data from the Institute of Labour and Manpower Studies (1984) illustrates this demand with the increase from 36,035 Filipino workers deployed in 1975 to 282,506 in 1981. Accompanying this labour deployment where a notable flow of Filipinos, notably businessmen, professionals and political activists seeking refuge and asylum in other countries in the two decades, from 1965 to 1986, of the Marcos regime (Garchitorena 2007), a period when the Philippines was placed under Martial Law to its end via Proclamation 2045 signed on January 17, 1981 (Youngblood 1982).

Contemporary patterns of Filipino migration have however changed from the formative years, characterized by two particular shifts. The first is what Garchitorena (2007, 3) states as the "feminization of migration" which has since the 1980s been marked by an increase in the number of Filipino women working abroad and currently accounting for 70% of the overall oversea Filipino worker population and characterized by distinct labour and destination patterns: nurses, midwives and care-givers heading to North America and Europe; domestic helpers to Hong Kong, Italy, the United Kingdom, Spain and other countries in Asia and Europe; and entertainers to Japan (Garchitorena 2007). The second contemporary shift is attributed to the increasing numbers of second and third generation migrants who are actively recruited by their overseas-based relatives in a phenomenon referred to as "chain migration" (Garchitorena (2007, 3). Such a form of migration tends to be prompted in large part by a migrant's wish to "(1) add a second income to help a family meet their financial needs; and to (2) reduce the loneliness they feel in a foreign land" (Anonuevo and Anonuevo 2002).

Despite successive administrative changes after the Marcos regime, Garchitorena (2007) notes that the movement of Filipino workers to other countries continued, if not accelerated, evidenced by data that shows an average of 30,000 Filipino migrant workers sent a month from 1984 to 1990, growing to more than 50,000 a month from 1991 to 1995, and to 60,000 a month from 1996 to 2001 (Garchitorena 2007). The numbers have significantly grown with current estimates indicating 2.3 OFS who worked abroad at any time during the period April to September 2017 with females making the larger proportion (53.7%) compared to males (46.7%) over this period (PSA 2018). Unlike the previous administration, the Benigno Aquino administration on its part set out a different agenda articulating instead for a policy which sought a transformational approach to Filipino migration away from the commonly held perception of being an "export commodity and means of earning foreign exchange". This policy, dubbed the "*16-point Social Contract with the Filipino People*" (NEDA 2011, 398), in part sought to mitigate against the "perils that many Filipino migrants are confronted with in the pursuit of a better quality of life" by offering an alternative option through more localized, inclusive and sustainable development efforts to be achieved by generating local employment and managing migration "in order to reduce costs, maximize benefits, and promote the rights, welfare and well-being of overseas Filipinos and their families" (Nicolas 2016, 34). The current Rodrigo Duterte administration has a mixed outlook on the OFW phenomenon, supporting it on the one hand, such as launching of an OFW Bank in 2018 meant to facilitate financial and remittance services "tailored to the needs of OFWs and their families" (Valente 2018), and in the same breath supporting endeavours that would induce the return of the OFWs back to the Philippines (Avendano 2018).

Despite political ambiguities on the Filipino diaspora, especially its OFWs, key characteristics of contemporary Philippine migration are characterized by "its massive scale and broad geographical spread, intergenerational character, diverse profile and varied migrant experiences overseas" (IOM 2013). Given its history, the trend in Filipino migration characterizes the Philippines as "a country with a long and vibrant history of emigration and the Filipino people considered the country's most valuable export" (Garchitorena 2007, 1). The country as a consequence has carved out its niche and in effect established migration corridors from the Philippines to other destination in North America, the Middle East, East and Southeast Asia, Oceania and Europe, in effect "making the Filipino presence, skills, talents and services truly global" (Nicolas 2016). This is more evident with support from the most recent data from the Commission on Filipino Oversea (CFO 2015) where by region of destination, the Americas have the most number of Filipinos (42%) followed by Western Asia (Middle East) at 24.5%, then East and South Asia (16.3%). Other regions of significant Filipino presence include Europe (8.5%) and Oceania (6.3%) of the same total estimate prior noted (Commission on Filipinos Overseas 2015).[1] Africa in general accommodates fewer Filipinos than other regions compared with only 0.53% of the 10 million estimated Filipinos abroad (CFO 2015). Despite the varied statistics, the Philippine view of the diaspora is collectively singularly one, as "inclusive and encompassing, recognizing a plural world of nations where Filipinos roam" (Aguilar 2015, 457).

Diaspora and OFW Remittances: The Goose that Lays the Golden Eggs

According to the World Bank (2015), remittances sent home by international migrants from developing countries are estimated to have risen to US$432 billion in 2015. India was the largest remittance-receiving country, with an estimated $69 billion in 2015, followed by China ($64

[1] Estimates does not include sea-based workers.

billion), and then the Philippines ($28 billion) (Nicolas 2016). In 2014, India, China, Philippines and Mexico were the top recipients of migrant remittances (World Bank 2015). Current estimates indicate that the total remittance sent by Filipino OFWs during the period April to September 2017 was estimated at $4 billion of which cash sent home comprised the largest proportion of this estimate ($3 billion), followed by cash brought home ($1 billion), and about $100 million through in-kind remittances (PSA 2018). Such remittances were sent mostly through banks followed by either local agencies, door-to-door deliveries, friends or co-workers or through other means (PSA 2018).

Ergo, the International Organization for Migration and Small Enterprises Research and Development Foundation (IOM) report that remittances made by overseas Filipinos tend to contribute to the development in the country in the following ways: (i) improvement in the quality of life of Filipino families and households through increased consumption, savings and investments, tis helping mitigate against poverty at the micro level; (ii) providing opportunities for migrants and their families to save, thus improving local capital markets; (iii) investing remittances in assets such as houses and land; (iv) spurring creation of micro-enterprises by families left behind; (v) putting up additional resources into education an heath; (vi) increase consumer ad demand for goods and services; and (viii) providing for safety nets such as social, cultural and infrastructure development projects (IOM 2012). Nicolas (2016, 35) in this regard opines that "remittances are the most visible contributions made by overseas Filipinos".

Various reports acknowledge the growing enthusiasm for the role that international migrants play in the development of their home countries. Hickey (2016, 681), on his part, contends that supranational development organizations, national governments, non-governmental organizations and other civil-society institutions "have [in effect] scrambled to assess the linkages between 'migration and development' and to implement new policies to facilitate transnational remittances and promote development". In the same contention as outlined by Hickey (2016), the Philippine State actively pursues this function by establishing agencies responsible for Filipinos going overseas notable of which is the Commission on Filipino Overseas (CFO). Established on 16th June 1980 through the Batas Pambansa 79 Act, the CFO is a government agency under the aegis of the Office of the President "tasked to promote and uphold the interests, rights and welfare of Overseas Filipinos, and strengthen their ties with the Motherland" (3rd Global Mayoral Forum 2017). The evolution of the agency is traced to the Presidential Decree No. 442 on the Labour Code of the Philippines of 1974 that initially created the Overseas Employment Development Board (OEDB). The Decree sought to set up a systematic programme for the overseas employment of Filipino workers, as well as to register and monitor emigrants leaving for various countries. Given a steady increase in migration resulted in the establishment of the Office of Emigrant Affairs (OEA) via Presidential Decree No. 1412 in June 1978, an independent unit constituting of local support network solely for Filipino emigrants which was subsequently replaced by the CFO in 1980 (3rd Global Mayoral Forum 2017).

Diaspora Engagement: The Crux of the Matter

Nicolas (2016, 36) observes that states are developing innovative strategies in efforts at taping migration's lucrative contributions to a country's development agenda. The author notes that these are done by actively engaging their diasporas who "have maintained strong sentimental and material links with their countries of origin". In the Philippine context, the roots of attempts at engaging the Filipino diaspora can be traced to the 1974 Presidential Decree No. 442 by President Marcos, when in addition to creating agencies specific to Filipino manpower deployment specifically the Overseas Employment Development Board and the National Seamen Board, established as well the Office of Emigrant Affairs housed in the Department of Labour, one that was deemed to be in line with the national policy of maintaining "close ties with Filipino migrant communities and promote their welfare as well as establish a data bank in aid of national manpower policy formulation" (POEA 2005). As articulated in Article 19, the Office was aimed at promoting the well-being of Filipino emigrants and aide in this process their close linkage with the homeland through various services deemed appropriate (POEA 2005). At present, the onus falls under the CFO, which as the lead State agency, has a number of initiatives geared towards promoting Filipino Diasporic engagement. A notable initiative of the CFO in this regard is the *BaLinkBayan programme* launched in 2013 serving as one-stop online portal offering an integrated platform for overseas Filipinos to re-engage with the Philippines through the Diaspora to Dialogue (D2D) programme (Nicolas 2016, 40). Another is the *Overseas Filipinos Remittances for Development* (or OF-RED) project operating at the sub-national level, and with support from the local government units (LGUs), facilitate the use of remittances for economic purposes such as job-creating investments and for micro-financing social enterprises (Nicolas 2016). At the national level, a multi-stakeholder forum dubbed the "Remittance for Development Council (ReDC)" offers consultative services for remittance-related stakeholders including the country's central bank, the *Bangko Sentral ng Pilipinas*, (BSP), private banks and non-banks, and to migration-related government agencies, migrant organizations, the academe and multilateral institutions (Nicolas 2016).

The Philippines being relatively advanced of other migrant-sending countries in the region of Southeast Asia such as Indonesia, in the development of initiatives that seek to engage with its Diaspora, Nicolas (2016) cautions that much still remains to be done "despite the plethora of initiatives" (Nicolas 2016, 46). A salient and welcoming feature in current policy discussions is the participation of new voices from various institutional actors, mostly non-governmental ones. This paper seeks therefore to contribute to this area of study by adding a view on the role of the private sector engagement with the Filipino Diaspora. Doing this will be in line with Nicolas' (2016, 46) suggestion that "the Philippines needs to go beyond the policy confines and spaces of traditional migration-related agencies and officials to involve development planners and other policy sectors and actors to expand the policy and programme options for the Filipino diaspora".

In this line of thought, the question begs on the contorting roles played by the *Blas F. Ople Center and Training Institute* and the *Global Filipino Diaspora Council* (GFDC) on best practices that their respective programmes play towards contributing to the general discourse on Migration-Development Nexus with respect to the *engagement* of the Filipino Diaspora.

The Global Filipino Diaspora Council (GFDC)

In September 2011, the CFO convened the *1st Global Summit of Filipinos in the Diaspora* in Manila in an effort to bring together outstanding Filipinos and community leaders worldwide and launch the *Diaspora 2 Dialogue (D2D) programme* as "an initiative that provides a framework and strategies for engaging the Philippine Diaspora in the development of 10 areas of interventions" (CFO 2016). Table 1 illustrates these interventions.

Table 1 Diaspora 2 Dialogue (D2D) Programme

Programme	*Purpose*
Alay Dunong Program (Brain Gain Program)	Seeks to systematize and strengthen the skills and technology exchange & transfer between Overseas Filipinos (OFs) and the Philippines with initial focus in the fields of science and technology, and engineering
Lingkod sa Kapwa Pilipino (LINKAPIL)	Aims at coordinating and facilitating the transfer of material and financial forms of assistance from OFs to development projects in the Philippines
Arts and Culture Exchange (ACE)	Aimed at promoting and facilitating the exchange of artistic and cultural products and workers between the country and Filipino communities abroad
Global Legal Assistance and Advocacy (GLAAD)	Mobilizing the support of both overseas and local Filipino lawyers in order to provide legal assistance and advice to OFs in distress, as well as advocate for laws, policies and programmes that empower Filipinos both in their home and host countries
Balik-Turo (Teach-Share) and Education Exchanges program	A continuous professional and vocational education exchange programme aiming for the return of overseas members of the academe in order to strengthen and enhance educational programmes and services in partner schools in the Philippines
Medical Mission Coordination (MMC)	Facilitates the conduct of medical missions sponsored and organized by OFs on medical interventions and in aiding the identification of medical need and impacts of these missions
Business Advisory Circle (BAC)	A matching and referral programme where business experts provide advice to OFs in setting up and sustaining their businesses
Return and Reintegration (R&R)	Focuses on providing returning OFs and retirees with updated information on services for successful reintegration into local life in the Philippines

(continued)

Table 1 (continued)

Programme	*Purpose*
Diaspora Investment (DI) program	Seeks new financial instruments and services supporting OFs such as rechannelling of remittances towards local and national development
Tourism Initiatives (TI)	Focuses on supporting Filipino overseas tourism and investment in local small-scale tourism enterprises

Source CFO (2016)

A major outcome of this Summit was the decision to form the *Global Filipino Diaspora Council* (GFDC) as the pioneering global network of overseas Filipinos (Nicolas 2016, 40). In February 2013, the GFDC, with support from the CFO and the US-based *US Pinoys for Good Governance* (USP4GG), convened the *2nd Global Summit of Filipinos in the Diaspora summit* aimed at tracking the progress of initiative mooted at the 1st Summit while highlighting best practices of the *Diaspora 2 Dialogue (D2D) programme*. A 3rd Summit was held in February 2015 where the GFDC in partnership again with CFO, convened the *3rd Global Summit* again in Manila which carried the theme: "Vision and Action for the Diaspora, 2015 and Beyond" (Nicolas 2016). Key outcome of the Summit was developing a strategic action plan for the Filipino diaspora per region of their settlement (Nicolas 2016, 40). This plan seeks to provide an avenue for continued networking and convergence of respective regional Filipino Diaspora Councils "in order to craft a common vision and development action plan for the Filipino Diaspora communities for 2015 and beyond" (Rodis 2015). Table 2 sums the key outcomes of the 3rd Summit relevant to the discussion on diaspora engagement.

It is the intention of the GFDC to be a relevant civil society in organizing Filipino diaspora conferences and councils in all continents where the Filipino is present (Rodis 2015). This expectation is inspired by what Rodis (2015) acknowledges is "the 'Indian model' where the Indian government views Diaspora not just as remittance senders but as an important bridge to collaboratively access knowledge, expertise, resources, and markets for the development of the country of origin" (Rodis 2015). The "Indian model", alluded to by Rodis (2015), can be referenced to the work of the Global Organization of People of Indian

Table 2 Outcome of the 3rd Global summit of Filipinos in the diaspora, 2015

Initiatives	*Activities*
The US Pinoys for Good Governance (USP4GG)	(a) Influencing Philippine domestic policy (lobbying for the passage of the Dual Citizenship Law and the Overseas Absentee Voting Act in 2003 and the removers of the "affidavit of return" provision in the Amended Overseas Voting Law in 2013) (b) Exercising domestic political rights within their host country, the US (election of more Filipino-Americans, influencing the passage of bills that provided citizenship to WWII veteran and the benefits they are entitled to) (c) Getting involved at the international level, on an independent capacity, to uphold the sovereignty of the Philippines (protesting against the threat of China)
European Network of Filipino Diaspora (ENFID)	• Aims towards changing the less desirable behaviour of Overseas Filipinos (e.g. gossiping & rumour mongering, the showbiz star mentality) to enhance Filipino reputation as a people, to engender better community relations and to serve as good examples to the 2nd and 3rd generations

Initiatives	*Activities*
The Middle East Network of Filipino Diaspora (MENFiD)	• Encourage Philippine and Qatar companies & business persons to establish operations in the Philippines and to promote entrepreneurship among Filipinos in Qatar. Through PINOYWISE, it provides financial literacy & addresses barriers to reintegration • Through the Foundation of Filipino Workers Worldwide (FFWW), it plans to create an OFW bank & hospital. It has also coordinated with the Qatar government for assistance towards Typhoon Yolanda victims

(continued)

Table 2 (continued)

Initiatives	*Activities*
Private sector Initiatives	1. The Medical City's world-class health care was presented where "patients are at the center stage and service delivery is of greater worth" 2. Efforts by the US Medicare Philippines in pushing for the portability of medicare benefits to the Philippines. It has created a task force that includes various stakeholders that is conducting a feasibility study of top Philippine hospitals paid by Medicare Insurance funded by the Philippine Retirement Authority and plans to propose amendments of the Medicare law for long-term care in the Philippines 3. Efforts by Science and Technology Action Networks for Development (STAND) to provide a global network for Filipino professionals in the Philippines and abroad to raise the visibility of their technical expertise in various fields such as health, business, agriculture, & the environment. It also recognizes the important role of information technology in science
Successful & innovative diaspora engagements	

Initiatives	*Activities*
Diaspora Investment Alliance (DIA) at the Aspen Institute, Washington DC	• In partnership with CFO, created the **Philippine Philanthropic Fund** to attract donations from the Filipino Diaspora which are tax deductible and which can be used for grants to NGOs & investments in social impact or enterprise development in the Philippines. An NGO Transparency Portal will provide information for diaspora interested in supporting Philippine NGOs
IdeaSpace	• Looks for promising Science & Technology ideas from the Philippines & Southeast Asia to support initial commercialization efforts, provides capital and expertise for them, and provides office space in Makati & housing for non-Metro Manila incubates • Its next big dream is to have the first Filipino-led Science & Technology Office in Silicon Valley
Kalibrr, Inc	• Uses the internet and digital communication for various business-related activities from connecting to pharmaceutical companies to pharmacies, to helping jobseekers look for jobs & companies to source workers to helping startup business ventures, to providing education-related solutions and addressing various social issues
Western Union Foundation	• Going beyond transfer of money to channelling resources to provide access to health & wealth creation, providing funds for business enterprises as well as providing financial literacy. It has provided funding for Peso Sense, the Philippine Financial Freedom Campaign of the UNDP and the CFO to improve the financial literacy of Overseas Filipinos and their beneficiaries

Source Go (2015)

Origin (GOPIO) which was founded in 1989 in New York, USA during the 1st Global Convention of People of Indian Origin. According to GOPIO (2017), the initial thrust of the Organization was fighting human rights violation of People of Indian Origin (PIOs). However with a relatively steady improvement in this area, the shift is now on the pooling of resources, both financial and professional, for the benefit of PIOs, in both the countries they come from and in India. Like GFDC, GOPIO International is a non-partisan and apolitical entity that envisions to "bring people of Indian Origin in the World together [in order] to promote their professional, cultural and social interests with a view to realizing their maximum potential in the service of Global peace, prosperity, happiness and compassion" (GOPIO 2017). To bring to fore PIOs around the world, GOPIO International has Chapters in Africa, Americas, Asia Pacific, the Middle East and Europe (GOPIO 2017), something the GFDC would like to emulate in terms of reach so as to bridge the capitals of its Filipino Diaspora which Rodis (2015) notes is depended on two conditions, firstly the ability of the Filipino diaspora to develop and project a coherent and progressive identity, and secondly, the capacity of the home country to establish sustainable, symbiotic and mutually rewarding conditions to further spur active diasporic engagement (Rodis 2015).

The Blas F. Ople Center and Training Institute

The *Blas F. Ople Center and Training Institute* (henceforth the Ople Center) refers to itself as "a non-profit organization that handles labour and migration concerns and develops programmes to empower the OFW and their families" (Blas F. Ople Center 2017). The Ople Center's mandate is to serve as a link between families of overseas Filipino workers (OFWs) and government agencies or other agencies to mitigate against problems arising from the migrant labour phenomenon. Key activities of the Ople Center include awareness programmes on events, policies and issues related to overseas employment, empowerment training programmes for OFW families, advocacy against human trafficking and illegal recruitment and lobbying for beneficial policies, programmes and legislation for OFW (Blas F. Ople Center 2017).

According to an email sent to me on the 20th of December, 2016 by Susan Ople, the youngest daughter of the late Labour and Foreign Affairs, the following are example of best practices of the Organization in helping distressed OFWs around the world as part of its diasporic engagement initiatives:

1. **Use of social media**. When a civil war broke out in Libya and the Department of Foreign Affairs (DFA) raised the Alert Level to #4, meaning mandatory evacuation for all OFWs, the Ople Center used Facebook to keep in touch with distressed workers. As the Philippine Embassy in Tripoli had no Facebook Account or a social media officer and the roads going to and from it were manned by several militia checkpoints, it was hard for the Filipino workers to get in touch with the Embassy. Ergo via social media, the Ople Center was able to cascade information to them regarding pick-up points and time when embassy personnel would be at various locations to bring the workers to safety. Ever since this experience, the Ople Center has been advocating for the government to assign social media officers in Philippine overseas embassies where there are a large number of OFWs. This will soon become a reality because the 2017 budget of the Overseas Workers' Welfare Administration (OWWA) provides for the creation of a social media unit within the agency and availing social media training for all welfare officers to be assigned abroad.
2. **Entering into strategic partnerships**. Because of the scarcity of grant money, the Ople Center enters into twining partnerships. For example, the Center had previous partnerships with Google Philippines and Microsoft Philippines to make Filipino OFWs more aware about their apps and technology as a whole, and its potential for more learning opportunities. There is currently an ongoing partnership with I-Remit, a non-banking remittance company that aided the Centre in conducting writing therapy workshops for mainly overseas domestic workers in Hong Kong.
3. **Fighting human trafficking**. The Ople Center works closely with the Department of Justice, the Bureau of Immigration, the Department of Labour and Employment and the Department of Social Welfare and Development to counter human trafficking of Filipino workers abroad. One of the best practices that can be emulated is the inclusion of the Center in the Inter-Agency Council Against Trafficking (IACAT) co-chaired by the Secretary of Justice and the Secretary of Social Welfare and Development. The Ople Center was appointed by the Office of the President for three terms as the lead NGO Representative of the Migrant Workers' Sector. As the NGO

representative, the Ople Center is able to work closely with government agencies in crafting strategies to combat human trafficking involving Filipino workers abroad.

From the two cases described, it can be postulated that these are in agreement with Nicolas' (2016, 45) assertion that "civil society organizations and the private sector have developed innovative ways to engage with and mobilize diaspora remittances and resources, especially for local economic development". The two cases in discussion are illustrative of scholarly discussions and debates on the discourse on "migration-development nexus" which Bohning (2009) outlines as occurring between those advocating for the role of human rights in enhancing the potential of migration and development while focusing on the economic dynamics of such circular. Despite their divergent focus, private organizations in general play vital roles in organizing stakeholders in a bid to forge networks and or tackle issues affecting Filipinos globally. What remains to be empirically determined, however, is the methodology employed by such "representative organizations" in their criteria for selecting members as part of their Forums that "represent" and "engaged" their respective Diaspora. In addition, there is a need to evaluate the effects and or impacts of both economic and socio-cultural remittances and or interventions to sustainable "development".

Discussion

The roles played by private entities with regard to engaging Filipino diaspora are indeed well summed by Nicolas (2016, 45) who notes the following inherent innovative practices involved in the discourse on Diasporic engagements for development: "(i) financial literacy and entrepreneurship training for migrants and their families back home; (ii) individual and collective savings mobilization schemes targeting the migrants themselves and their families back home including their children; (iii) investments in products of already established industries or programmes of micro-finance institutions; (v) twinning arrangements with hometowns in the Philippines and areas in countries of destination; (vi) linking migrant capital with social enterprises and (vii) business and enterprise development services for Filipino migrants".

Nicolas' (2016) outlined gains align with "a simultaneous growth of scholarship" on the discourse on "migration-development nexus" (Hickey

2016; Kapur 2004) in line with the following fundamental elements as espoused by Geiger and Pécoud (2013, 369): "(a) the central role played by migrants and diasporas in the development of their countries of origin through remittances or return in particular; (b) the potential impact on world development of even a small liberalization of labour mobility; (c) the complementary situation of sending and receiving countries, which need emigration and immigration to ensure growth; and (d) the consequent attractiveness of temporary and circulatory labour migration schemes as an ideal compromise, and more generally, as an example of 'sound', 'balanced' or 'well-managed' immigration policies".

Scholars however caution against such an enchanting discourse on the premise that espoused developmental gains from migration is likely to foster (rather than reduce) migration and thus add to resultant social stresses (Martin and Taylor 1996; Massey 1991). This stance is contrary to what many policymakers still contend that trends in local development would result into less migration, reflecting what Geiger and Pécoud (2013) note is "the 'older' idea of development and migration as a zero–sum game". Divergent views still persist in discourse regarding financial versus social nature of remittances (de Haas 2007; Levitt 1998). This is coupled with the lack of an adequately informed and expanded knowledge base in many developing countries, such as the Philippines, to derive prescriptive recommendations about policies that are likely to enable them make best uses of various forms of diasporic remittances.

Ergo, advise is given on the need to always keep in mind "the deep and structural socio-economic and political issues at stake behind the enthusiasm in migration and development where the focus should lie less on whether migration will foster development, or whether political initiatives in this field will last or not, and more instead on the implicit assumptions behind the interest in the 'migration-development nexus', and in what they tell us of the determinants of both development and migration policy" (Geiger and Pécoud 2013, 373).

Conclusion

As different countries of the world engage their diasporic communities for different purposes, they nonetheless commonly maintain their engagement for the mutual benefits of all; home countries, the Diaspora and the host countries. A salient and welcoming feature in current policy discussions on diaspora engagement is the participation of new voices from various institutional actors, mostly non-governmental ones as exemplified in the cases of the *Blas F. Ople Center and Training Institute*, and the *Global Filipino Diaspora Council* (GFDC) on programmes that contribute to the general discourse on Migration-Development Nexus within the context of engaging the Filipino diaspora. As prior mentioned, it remains to be empirically determined on the methodology employed in the criteria for selecting members as part of representative Diaspora organizations that seek to "represent" and "engaged" their respective Diaspora. In addition, there is a need to critically evaluate the effects and or impacts of both economic and socio-cultural remittances, and or interventions, to sustainable "development" as part of the discourse on Migration-Development nexus. What remains therefore is the challenge to what Geiger and Pécoud (2013, 373) note, is the unveiling of biases that characterize the migration and development nexus and unveil the dominant biases that characterize this discourse, while equally paying attention to its consequences in moulding the way migration and development are thought about-and therefore governed by both State and non-State actors alike.

References

Aguilar, Filomeno V. Jr. 2015. Is the Filipino Diaspora a Diaspora? Critical Asian Studies 47(3): 440–461.

Anonuevo, Estrella Dizon and Anonuevo, Augustus T. 2002. Coming Home - Women, Migration and Reintegration. Manila: Balikbayani Foundation, Inc. and Atikha Overseas Workers and Communities Initiative, Inc.

Avendano, Christine. 2018. Duterte urges 260,000 OFWs in Kuwait to come home. Philippine Daily Inquirer. April 28 208. Accessed 30 April 2018. http://globalnation.inquirer.net/166306/rodrigo-duterte-ofws-in-kuwait-homecoming#ixzz5EA2A8P3C.

Blas F. Ople Policy Center & Training Institute. 2017. "Blas F. Ople Policy Center and Training Institute." Accessed December 30 2016. http://www.blasoplecenter.com/about-us/.

Bohning, Roger. 2009. Getting a handle on the migration rights-development nexus. International Migration Review 43(3): 652–670.

Commission of Filipino Overseas (CFO). 2016. "About the Diaspora to Development or D2D Program." Accessed December 20 2016.http://www.cfo.gov.ph/15-aboutcfo/programs-and-projects/2997-about-the-diaspora-to-development-or-d2d-program.html.

Commission on Filipinos Overseas (CFO). 2015. “Stock Estimates of Filipinos Abroad. Commission on Filipinos Overseas.” Accessed December 20 2016. http://www.cfo.gov.ph/index.php?option=com_content&view=article&id=1340:stock-estimate-of-overseas-filipinos&catid=134:statisticsstock-estimate&Itemid=814.

De Haas, Hein. 2007. Remittances, Migration and Social Development. A Conceptual Review of the Literature. Social Policy and Development Programme Paper 34. Geneva: United Nations Research Institute for Social Development (UNRISD).

Garchitorena, Victoria P. 2007. Diaspora Philanthropy: The Philippine Experience. Boston MA: The Philanthropic Initiative and the Global Equity Initiative.

Geiger, Martin and Pécoud, Antoine. 2013. Migration, development and the ‘migration and development nexus’. Population, Space and Place 19 (4): 369–374.

Global Organization of People of Indian Origin (GOPIO). 2017. About. Accessed January 30 2017. http://www.go-pio.org/about/.

Go, Stella. 2015. Synthesis of the 3rd Global Summit of Filipinos in the Diaspora: Vision & Action for the Filipino Diaspora 2015 & Beyond’. 3rd Global Summit of Filipinos in the Diaspora. Manila. 25 –27 February 2015. Accessed January 3 2017. http://3gs.cfo.gov.ph/download/day-2-plenary/workshop-output/Synthesis3rdGlobalSummit.pdf.

Hickey, Maureen. 2016. Modernisation, Migration, and Mobilisation: Relinking Internal and International Migrations in the ‘Migration and Development Nexus’. *Population, Space and Place* 22, 681–692.

Institute of Labour and Manpower Studies.1984. Working Abroad: The Socio Economic Consequences of Contract Labour Migration in the Philippines. Manila: Institute of Labour and Manpower Studies, Ministry of Labour and Employment.

International Organization of Migration. IOM. 2013. Country migration report: The Philippines 2013. Makati: IOM and Scalabrini Migration Center.

International Organization of Migration (IOM). 2012. Channelling remittances for development: A study towards creating model mechanisms. Makati: IOM and Small Enterprises Research and Development Foundation.

Kapur Devesh. 2004. Remittances: The New Development Mantra? UNCTAD G-24 Discussion Paper Series 29. New York: UNCTAD.

Levitt Peggy. 1998. Social remittances: migration driven local-level forms of cultural diffusion. International Migration Review 32(4): 926–948.

Martin, Philip, L. and Taylor, Edward, J. 1996. “The Anatomy of a Migration Hump.” In Development Strategy, Employment, and Migration: Insights from Models, edited by Edward Taylor, 43–62. Paris: OECD Development Centre.

Massey, Douglas, S. 1991. Economic Development and International Migration in Comparative Perspective. Population and Development Review 14 (3): 383–413.
National Economic Development Authority (NEDA). 2011. "Philippine development plan 2011–2016." National Economic Development Authority. Kiel: Kiel Institute for the World Economy.
Nicolas, Imelda, M. 2016. "Engaging the Global Filipino Diaspora: Achieving Inclusive Growth." In Diasporas, Development and Governance. Global Migration Issues 5, edited by Chikanda, Abel, Crush, Jonathan, Walton-Roberts, Margaret, 33–47. Heidelberg: Springer International Publishing.
Philippine Statistical Authority (PSA). 2018. Total Number of OFWs Estimated at 2.3 Million: Results from the 2017 Survey on Overseas Filipinos. Philippine Statistical Authority. Accessed April 30 2018. https://psa.gov.ph/content/total-number-ofws-estimated-23-million-results-2017-survey-overseas-filipinos.
Philippine Overseas Employment Administration (POEA). 2005. "Labour Code Provision on Overseas Employment: The Labour Code of the Philippine Presidential Dercree No. 442 as Amended."Accessed January 5 2017. http://www.poea.gov.ph/laws&rules/files/Labor%20Code%20of%20the%20Philippines%20Provisions%20on%20Overseas%20Employment).html.
Ramos, Ricardo. 2013. "Laments on the Filipino Diaspora." Philippine Daily Inquirer, August 10. http://globalnation.inquirer.net/82987/laments-on-the-filipino-diaspora
Rodis, Rodel. 2015. "Vision and action for the Filipino diaspora." Philippine Daily Inquirer. March 10. https://globalnation.inquirer.net/119428/vision-and-action-for-the-filipino-diaspora#ixzz3U2UbgTi1.
Sicam, Paulynn Paredes 2003. Philippine Migration Journalism: A Practical Handbook. Quezon City: Institute on Church and Social Issues, OFW Journalism Consortium.
Valente Catherine. (2018). Duterte Opens OFW bank. The Manila Times. January 18. Accessed 30 April 2018. https://www.manilatimes.net/DUTERTE-OPENS-OFW-BANK/375119/.
Vera Cruz, Philip 1994. A Personal History of Filipino Migrants and the Farm Workers Movement. UCLA Labour Center: Institute of Industrial Relations.
World Bank. 2015. Migration and Remittances: Recent Developments and Outlook. Migration and Development Brief 24. May. Washington D.C.: International Bank for Reconstruction and Development /The World Bank.
World Bank. 2016. Migration and Remittances: Recent Developments and Outlook. Migration and Development Brief 26. April. Washington D.C.: International Bank for Reconstruction and Development /The World Bank.
Youngblood, Robert. 1982. The Philippines in 1981: From "New Society" to "New Republic". Asian Survey 22 (2): 226–235.

3rd Global Mayoral Forum. 2017. "3rd Global Mayoral Forum on Human Mobility, *Migration and Development." Accessed December 24 2016.* http://3rdglobalmayoralforum.cfo.gov.ph/about-us/partners/commission-on-filipinos-overseas.

The Mexican Diaspora in the US and the 3 × 1 Program as Transnational Community Development: Lessons and Challenges

Rodolfo García Zamora

Introduction

Since the end of the seventies (from the past century) Mexican migration to the US grew extraordinarily from 800 thousands to 12 million in 2008, and this takes place proportionately to the depth of the Mexican economic crisis in 1976, 1982, 1994 and the economic stagnation from 2000 to 2008. Emigration's constant growth is fundamentally explained by those Mexican crises and a long period of positive economic growth in the US, which for four decades acted as an "escape valve" for the Mexican structural problems of employment, income and well-being (García 2009: 309).

Therefore, it is possible to observe that during Vicente Fox's term (2000–2006), 575 thousands emigrated each year, 3 million in total, provoking among other things the depopulation of hundreds of communities in at least 600 municipalities in the country with a significant migratory tradition, and registering then negative rates of demographic growth. Indeed, in that period, the number of women abandoning the

R. G. Zamora (✉)
Universidad Autónoma de Zacatecas, Zacatecas, Mexico
e-mail: rgarciaz@prodigy.net.mx

A. Dubey and S. Rajwar (eds.), *Diaspora in Development*,
https://doi.org/10.1007/978-981-96-6747-5_10

country in search of the American dream increased to be 45% of the total (Balboa 2007).

During the first year of Felipe Calderon's government, the growing tendency of emigration to the US remained constant according to the data of the National Population Council[1] for 2007, estimating that 679,611 Mexican established in the US (Garduño 2008). The World Bank confirms these tendencies (Milenio 2008) when they inform that in recent years 644,361 Mexicans left the country on average each year. Thus, Mexico has expelled 10% of its population.

However, during the summer of 2011, some papers about Mexican international migration to the US talked about a historical turning point in migratory tendencies in both countries, which is something that never happened before in the last forty years: a significant decrease of migration. According to the data from the PewHispanic Center, less than 100 thousands undocumented migrants went to the US in 2010 which is much lesser than the 525 thousands established annually in the US between 2000 and 2004, undocumented population in the US has remained at 11.2 million (El Universal 2011). This figure implies a decrease of one million migrants compared to the figure in 2007, when total undocumented migration reached the limit of 12 million. The impact of the economic crisis in the US and a stricter border control, are considered the main causes explaining the decrease. In spite of that, undocumented population is three times bigger than the one calculated in 1990 and within this population, the Mexican origin population remains the biggest one having 6.5 million of undocumented people.

This new tendency fueled a debate about migration's nature: either circumstantial or permanent, this was known as the "zero migration" debate. What is relevant in this debate is Mexico's extreme fragility, severely affected by the economic crisis of the neighbour country, which questions three decades of structural reforms that failed to eradicate the causes of migration and failed to get rid of the high dependence upon the economic cycle of the US. Thus, this country cannot play anymore the role of an "escape valve" for Mexico and hence, it is necessary to reorient the national economic model towards employment, income and the well-being of its population (García-Zamora 2012a).

[1] NB. The original name in spanish is: Consejo Nacional de Población.

According to Alba (2008) the future tendencies of the factors explaining migratory patterns between Mexico and the US suggest the continuity of the powerful migratory pressures in the mid and the long run. Concerning the US, the majority of the observers agree expecting that structural demand for transnational migrants will remain constant in the next future. For that reason, if Mexico's economic growth stagnation and weakness persist and no solution is found to overcome its low capacity to absorb labour force (in spite of the high degree of integration with the US economy), then the strong migratory pressures will prevail. It has been estimated that even a moderated and sustained economic growth rate of 5% annually in Mexico will not be enough to contain the number of emigrants in the first decades of the twenty-first century (Tuirán et al. 2000). When thinking about the next future the persistence of huge gaps between the two countries will keep affecting the migratory results in the context of the growing interdependence and integration of Mexico and the US.

Therefore, in the context described above, and from a developmental perspective, Alba (2008) considers that Mexico has much more to do and it should adopt an explicit policy to "absorb" its labour force and citizens and deactivate a deeply rooted migratory mentality. In general terms, Mexico's development experience in the twentieth century shows that all different periods and strategies implied relatively massive migratory flows. Thus, there must be quantitative and qualitative changes in Mexico's economic policy to accelerate growth rates and to close regional and social gaps. It is an imperative to improve population's quality of life and improve employment opportunities in Mexico and not only in the poorer areas of the country.

According to the estimation based on the U.S. Census Bureau 1990–2016, in 2015 it is possible to see a progressive increase in international migration to the US with a stock of 12.2 million Mexicans, 5.8 million of undocumented migrants and a contribution of 24.7 billion US dollars as remittances, which turns out to be 30% higher than the total amount of foreign currency obtained by oil exports. The income Mexican migrants send to Mexico is the main source of foreign currency. As the estimation shows, Mexican migration to the US remains at a constant level in recent years (2015–2016). In this new context Arroyo et al. (2010) identify as the main challenges for the Mexican government and society as the need to search an endogenous economic model and a policy able to generate conditions that would avoid another Mexican generation looking towards the north as the only alternative to social mobility. There should be local and regional development policies oriented to decrease the negative socio-economic impacts of a decrease in the volume of remittances and oriented to decrease the number of migrants. This should be done considering specially the pressure of return migration over labour markets along with young Mexicans seeking to integrate to the labour force, as they do not have the chance to migrate to the US, at least not in the relatively favourable conditions of the past.

In the same line, Escobar (2008) argues that development and migration-related policies in the next years should aim at three objectives:

(1) They should offer real incentives (employment, income, housing, education)
(2) They should facilitate the access to the whole set of basic patrimonial goods (work, education, health, support for the elder)
(3) They should be able to use instruments according to the size of the intended tasks. It is not possible to change millions of migrants into development agents in their towns and regions on the basis of good social programs but rather symbolic and with a small scale. Government should assign enough resources and should develop an economic policy at the national and regional levels, able to promote growth as needed.

The proposed economically active policies, oriented towards internal markets, to generate employment and to reduce poverty seem to get more relevance at the beginning of 2017 when Mexico is suffering again a deep socio-economic crisis. The growth rate of the Gross Domestic Product is 1.6% in 2016, the debt is as high as 9.3 billion pesos (it represents 50% of the GDP), the peso devaluation in four years is 40%, oil prices plunge, the value of gas imports from the US reach 17 billion of US dollars, 56 out of 121 million inhabitants are poor, 60% of the economically active population is in the informal economy, there are 7.8 million of young people not studying and not working, there are 100 thousands of murders and 30 thousands missing persons in a period of eleven years (Fernández-Vega 2017). In this context, and since the end of 2015, international migration of Mexicans to the US and their remittances become strategic issues regarding Mexico's dynamics and processes.

However, the compensating effect that international migration and remittances have over the deep socio-economic Mexican crisis, the estimates of international remittances which have continued from the year 1995 was threatened in November 2016 when Donald Trump wins the presidency of the US and reaffirms his campaign promises to radically modify the Free Trade Agreement with Mexico on the basis of a neo-protectionist perspective, punishing firms that would invest in Mexico and applying high rates to their exports which mainly correspond to the car industry (35%). Moreover, Trump has announced his intention to deport 5.8 million of undocumented migrants, to severely deal with remittances and to build a wall between the two countries forcing Mexico to pay for it. So, in the context of long term economic stagnation from 2000 to 2015, the fall of oil prices as well as of the federal government revenues, the growing of the public debt, unemployment, poverty and all kind of violence, all consequences are intensified by a national budget for 2017 decreased by 4.8 billion pesos, in which 500 billion pesos are assigned to pay public debt and moreover, sectors such as health, education, science, technology and culture suffer a reduction of 239 billion pesos.

In that context, Trump's speech at the investment ceremony as President of the US, about "regaining" border control and "strengthening" internal efforts obviously confirms his neo-protectionist attitude in the economic and commercial fields. It also confirms his intentions to repress and massively deport undocumented migrants, mostly Mexicans. Indeed, concerning the economic protectionism his Secretary of Commerce has announced that one of the first actions will be a revision of the NAFTA. To deal with it, the Mexican government has no alternative economic development project weather by sectors or by regions aiming to strengthen internal markets and employment. Instead, the Mexican government enables the former negotiators of the original NAFTA 1991–1993 to receive the new neo-protectionist terms from the North American Empire. A republican administration is radically changing the socio-economic and geopolitical relations between Mexico and the US, and specially the migratory system which comprises 36 million people with Mexican origin, 12,2 million Mexicans living permanently in that country and 5.8 million of undocumented migrants facing—as we said before—massive deportation in the next years.

The evaluation of Mexico's selected exports revenues and remittances from the period 1995–2016 based on Banxico data shows that in Mexico, the value of remittances is only below oil exports revenue and this confirms the relevance of such remittances for the country's stability. In such complex circumstances for both countries, Mexican migrant organizations living in the US, working for the last forty years promoting various philanthropic programs to invest in social development in thousands of communities, will be affected by the deep changes that Trump's administration may provoke inside the US and in Mexico's economy, society and politics. The next section summarizes the Mexican Diaspora in the US, its emergence, their community programs of transnational solidarity, its contribution and the challenges it faces now and the next years.

The Philanthropy of the 3 × 1 Program as Transnational Community Development

In 2012 we observed that during the previous 15 years there was a significant growth of international migration from Mexico to the US, reaching 12 million Mexicans settled in that country. There was also an extraordinary growth in remittances entering Mexico, reaching in 2008 the amount of 26 billion of US dollars. Along with that, there is a growing relevance of more than 900 Mexican migrant organizations in the US because of their financial contribution to more than 12 thousand social projects of basic infrastructure in their origin communities (García-Zamora 2012b: 97).

The relevance of those organizations is the result of a transnational organizational process originated in the sixties from the last century and that's the case of pioneering migrant organization from the central-western region of the country, including Zacatecas, Michoacán, Guanajuato and Jalisco having more than a hundred years of transnational migratory experience which may explain why they were promoting—five decades ago—social projects in their origin communities using only their collective remittances. Migrants have labelled this period as "zero dollars for one" because it worked with their own contributions as an instrument to finance transnational community projects and this has been taken as a background the way that in the eighties of the last century it was possible to negotiate with some municipalities in a rather informal way, to carry out new projects under the "one by one" modality, one dollar from migrant clubs and one dollar from the municipality. Afterwards, at the beginning of the nineties, the Program changed in Zacatecas and Guerrero to a "two by one" modality, one dollar from the migrants, and one dollar from the federal government and another dollar from the state government. In 1999 in Zacatecas the "three by one" Program was established when municipalities decided to contribute one dollar additionally considering the fact that they started to receive more resources from the federal government.

The importance of the 3 × 1 Program as a valuable experience of transnational philanthropy is recognized by the Mexican government to the point that in 2002 it decides to create the "3 by 1 Citizens' Initiative Program" under the Micro-regional Program of the Secretary of social development with a specific federal budget for the first time. This is because previously, governments had to negotiate with the Secretary the financing of social projects from migrants using resources from other programs already established such as the Temporal Employment Program that served as a platform to orient financial support. The creation of the 3 × 1 Program as a national program was a way to institute spontaneous transnational philanthropy from below and from outside in terms of an incipient transnational social policy (Fernández de Castro et al. 2006).

The growth of the 3 × 1 Program as national program was extraordinary. In 2002 there were 20 big Mexican migrant organizations in the US financing 942 projects. In 2010 there were one thousand migrant organizations in 40 states of the American Union, financing 2438 projects and in 2015 there were 1697 migrant organizations in 41 states financing 1949 projects. In 2015, Mexican migrant organizations contributed to the 3 × 1 Program with 22 million US dollars that were completed with other 22 million from the federal government and from the state and local governments. From 2002 to 2016 they financed 26 thousand social projects of basic infrastructure with a transnational community development perspective (Presidencia de la República 2016). The figures are based on the elaboration with data of 3 × 1 Program Projects from the Presidencia de la República: 4th Presidential Address between 2015 and 2016.

As we said before, Mexico-US migration explains the emergence and progress of the Mexican migrant philanthropic projects developed in their origin communities, whose long tradition dates back to more than hundred years ago. The neighbour countries have a common border of more than 3 thousand kilometers and migratory flows are massive reaching 36 million people of Mexican origin in the US in 2017. These three factors explain the emergence, progress and the growingly leading role at the international level of Mexican migrant organizations, especially concerning their contribution to well-being improvement in their origin communities through financial support of thousands of basic infrastructure social projects in association with the three Mexican government levels: federal, state and municipal.

In fact, and in Mexico in general, these organizations and their projects have the following positive impacts all over the country (Gracía Zamora 2006):

1. They promote transnational community organization.
2. They make possible negotiation of migrant organization and origin communities with all government levels.
3. They promote social infrastructure building in communities and regions that were traditionally set aside.
4. They create a transnational learning process for all actors involved in the 3 × 1 Program.
5. They support the emergence of a new transnational social actor: transnational communities as potential agents for local development.
6. They help to promote Mexico's transparency and accountability culture which is starting to grow in communities and municipalities where the projects take place.

However, migrants' social projects have faced problems and difficulties related with selection, design and execution. In that sense, Orozco (2004) highlights three main issues:

- Migrants' contributions are often bigger than public resources for social projects in the communities.
- There are many disagreements between migrant clubs and municipal and state governments about selection processes, expected quality and maintenance.
- Origin communities have organizational weaknesses limiting their ability to control investments, projects' quality and maintenance.

After many years studying Mexican-migrant organizations in the US as well as their projects, it is possible to underline five main lessons:

1. There is a significant potential to support development strategies. Migrant clubs actions in this area increase economic opportunities

in regions with high international migration. However, to understand clubs' motivations to participate in development promotion it is necessary to understand their practices, their visions and their projects, the way it could be possible to identify collaboration opportunities with them.

2. There is commitment with relevant development projects. They have the capacity to promote and support small-scale local development projects.
3. There is a high diversity among Mexican-migrant clubs and organizations with sharp differences between "mestizo" Federations with a long history (Zacatecas, Michoacán, Jalisco and Guanajuato) and more recent Federations with high indigenous participation (Oaxaca, Puebla, Guerrero and Yucatán). Even between Federations of the same state as Zacatecas, for instance, there are clear differences between organizations from southern California, Chicago and Texas in terms of their degree of integration, number of years, social and political participation, autonomy, alliances, experience in social and productive projects and in their relation with the three Mexican government levels.
4. There is collaboration with other social actors. As a result of each migrant club and federation experience there is a particular history and specific capacities for collaboration with other social organizations in Mexico and in the US.
5. There is the ability to identify community needs. Clubs and origin communities can play a key role as local development agents, elaborating the agenda with social and productive projects considered as a priority for local and regional economic development. Nevertheless, this requires to promote a better transnational community organization, integrating levels such as the institutional, economic, social, educational and encouraging transnational community leadership. This is one of the strategic challenges that Mexican migrant clubs and Federations face.

The former assessment gets more relevant as Federations of migrant clubs from Zacatecas, Michoacán, Oaxaca, Guerrero and others, are trying to move from social projects to productive micro-projects and towards local development models with a transnational perspective. In such context, 3 × 1 Program could become one additional action within a set of activities specifically oriented. Therefore, to assess the possibilities to achieve such a transition in the next years, it is very important to evaluate the impact of the 3 × 1 Program at the economic, political, civic, social and cultural levels.

In 2015, the total contribution made by migrant organizations and by the three levels of the Mexican government to the 3 × 1 Program was 88 million US dollars. If we compare that to the 24.7 billion US dollars that migrants sent as remittances, we can say it is too little. However, if we consider that those funds developed 26 thousand projects in states such as Zacatecas and Michoacán, we could see that social infrastructure projects such as water, drainage, electricity, urbanization, pavement, schools, clinics, community centers and—increasingly—scholarships, represent the main investment initiatives in most of the country's municipalities with high international migratory intensity. Those projects have significant multipliers effects on the local economy, on employment, on the demand of building supplies and have a direct impact on population's well-being. Taking into account the later, some scholars consider the 3 × 1 Program as a "Welfare" system that benefits migrants' relatives. Migrants in the US along with the government finance those community projects.

This experience has stimulated a tendency where majors themselves promote the creation of migrant clubs in the US as a mechanism to make greater community investments with the 3 × 1 Program. As a result, they obtain more credits in the political context.

Mexican state is obliged to provide the aforementioned projects to its population, nevertheless, before migrant organizations and before the creation of the 3 × 1 Program the origin communities where set aside from such services. For that reason, some interpret the Program as the equivalent to a transnational social policy promoted by the Mexican diaspora in the US, which in turn improves social well-being and strengthens the local productive and social complexes. In that way, it could be possible to move towards a new type of social and economic activities that would allow us to overcome the prevailing structural economic fragility that has been caused largely by international migration.

There is an additional economic impact and it has to do with the aforementioned interest of some Mexican migrant organizations such as Club Federations from Zacatecas and Michoacán to move away from basic infrastructure social projects to productive projects after many years promoting hundreds of social projects in their origin communities. As this is a great challenge, it has been called the "deadly step" (García Zamora

2003) because of its enormous complexity and because it confronts us with our own country's institutional weakness to promote productive investment with the absence of entrepreneurial culture among the Mexican diaspora in general. Despite such difficulties, both Federations have been struggling during the first decade of the twenty-first century waiting for the Social Development Secretary to stimulate specifically migrants' productive projects with the 3 × 1 Program. In this way, the Program has supported 429 productive projects from 2002 to 2009 and the numbers keep growing gradually, especially in the case of greenhouses to produce tomato, livestock production and agave and tequila production.

Productive projects within the 3 × 1 Program have a great relevance for origin communities and their migrant organizations in case of success as well as in case of failure as it entails a collective learning process that contributes to improve processes and outcomes. From a historical perspective of more than three decades, the 3 × 1 Program and the former versions (1 × 1 and 2 × 1) represent a major advancement as a transnational organizational effort to promote social development in Mexican origin communities. The key to success stems on migrant organization's solidarity with their origin communities, along with an association with three levels of the Mexican government.

Summarizing, considering that the 3 × 1 Program has developed more than 26 thousand community projects it is possible to affirm with no doubt that it has contributed positively to the well-being of migrants' origin communities. Nevertheless, the institutional change towards a productive projects phase requires assuming a qualitatively different logic. This is the logic of firms, market returns, efficiency and it implies technical consistency criteria to assess projects as well as their economic sustainability. In this context, it is necessary to elaborate a specific strategy for technical, organizational and financial support for Mexican migrants' productive projects. In order to do so, it can be possible to count as an important support the progress and contributions of the 3 × 1 Program with respect to the transnational community organization, the new spaces for negotiation, the transnational social networks, and even the many propositions made by the Advisory Council of the Institute for Mexicans

Abroad.[2] Which is constituted by the Foreign Relations Secretary and by 130 representatives of migrants in the US.

Moreover, it is worth to notice that pretending to blend solidarity with productive projects in the context of the current regulations and functioning of the 3 × 1 Program entails the risk of destroying migrants' common asset, transnational communities' and Mexico's heritage. This could happen when profit logic would break solidarity and philanthropic logic within social contexts characterized mostly by the absence of an entrepreneurial culture.

The already mentioned risk implies the possibility of losing a valuable effort made by the Mexican transnational community who took forty years of social learning, building new spaces to negotiate with the three Mexican government levels and creating the basics of a new type of public policy for migrants, their communities, their families and their projects. Therefore, the challenge stems on the need to clearly assess the feasibility to proceed to the phase of productive projects and transnationally focused local development under the 3 × 1 Program by means of a deep reform. Otherwise, it is necessary to assess if the Program should keep promoting transnational philanthropy projects and consequently, a new federal program should be created to support migrants' productive projects and transnational local development.

With this perspective, it becomes interesting to see that in January 29, 2009 the Social Development Secretary pretending to cope the socioeconomic impacts of the US's financial crisis on Mexico informs through the Federation's Official Journal[3] the creation of the Productive Projects Program for Asset Strengthening. This program consists of a Fund to finance individual or family productive projects inside the framework of the 3 × 1 Program for Migrants. According to the exposed arguments in the Official Journal (2009), this Fund results from the migrants' constant demand to generate productive opportunities in their origin communities, to be able to increase their assets and their families' assets and to offer jobs to the young people in order to reduce their migration needs. At this point, it is possible to notice that the members of the Advisory Council of the Institute for Mexicans Abroad already presented such demand in

[2] N.B. The original name is: Consejo Consultivo del Instituto de los Mexicanos en el Exterior.

[3] N.B. The original name is: Diario Oficial de la Federación.

many occasions as part of their recommendations to the Mexican government in previous years. Under this investment mechanism, the Social Development Secretary contributes with 50% of each project's investment and migrants contribute the rest. Those resources will support individual or family productive projects presented by migrants settled abroad. Projects will develop in Mexican territory and each project will have as federal contribution no more than 300 thousand Mexican pesos (30 thousand US dollars in 2009).

By the end of January 2010, the Institute for Mexicans Abroad (2010) informs a new program for productive investment for migrants named: "Fellow Countryman, invest in your Land",[4] promoted by the Agriculture Secretary and the Shared Risk Trusting Fund (SRTF). This Program pretends to strengthen migrants' economic potential and the productive use of remittances. To justify the new program, the following arguments were developed.

Economic growth and the development of the Mexican countryside, require resources to invest in projects that can generate employment and wealth. There is a great potential in the 21 billion US dollars remittances sent to Mexico in 2009 and in a "nostalgia market" consuming more than 24 billion US dollars yearly in the US. There is a need for government support and stimulus to use remittances productively. The main aim of the new program is to contribute to the transformation of remittances into productive resources by investment in migrants' and their families' businesses. The specific goals are:

(a) Stimulate employment in origin communities to reduce depopulation and emigration.
(b) Take advantage of the "nostalgia market" in the US.
(c) Overcome remittance dependence transforming it into a new relation of productive partners while strengthening intergenerational relations of migrants with Mexico.

This new Program tries to solve the extended demand of Mexican migrant organizations who are interested in a specific Program from the Federal Government to support productive investments in origin communities. The implementation of the Program represents a progress in itself;

[4] N.B. The original version is: "Paisano, invierte en tu tierra".

however, it is important to assess weaknesses and risks for migrant clubs and their projects in order to participate adequately in the new Program with positive outcomes.

As we have seen the Mexican diaspora promoting for more than three decades social projects in Mexico and trying in the last fifteen years to move towards productive projects, we are now able to identify their weaknesses in both activities. In general, there are low organizational levels and limited capabilities in both, origin communities and migrant organizations. This, together with the absence of an entrepreneurial culture in both sectors is preventing to take advantage of the investment possibilities and thus, it explains the low percentage of successful productive projects (5 to 8%) among all financed projects by the 3 × 1 Program.

The absence of training programs and micro-enterprises management capabilities and the lack of permanent technical support in origin communities explain, to some extent, the previously mentioned weaknesses. Moreover, all of this is the result of the lack of regional economic development policies since 1982 when Mexico adopted the neoliberal economic model. Indeed, this also reflects the lack of a real social development policy that should have an impact over the structural causes of marginalization and poverty, which are—ultimately—the causes for international migration. An additional factor is the absence of the entrepreneurial sector in the context of the migrants' diverse philanthropic initiatives and in their investment programs for productive projects.

The challenges for public policies on migration and development in Mexico and to move from migrant philanthropy to the transnational and social economy are:

1. The possibility to collaborate between the three government levels, origin communities, migrant organizations and the private sector in order to design regional and local development strategies.
2. The possibility to establish a real Social Development Policy, able to influence the structural causes of marginalization and poverty, linked to a strategy of Solidarity Economy. This has to do with the need to replicate and improve the experience of the Southern California Zacatecan Club Federation that created its own technical agency in Zacatecas—Mexico to promote migrants' and their families' social and productive projects (36 productive projects in 2012).

3. The possibility to design along with Mexican and American universities, transnational training and organizational programs for the origin communities and migrant organizations.
4. Design and implement a program to support micro and small enterprises for migrants and their families under the integral value chains perspective.
5. Design and establish micro communitarian banks to encourage savings, credit, insurance and investment for migrants and their families.
6. Stimulate greater collaboration with American and Mexican universities in terms of technical advisory, university extension, social service and distant education with origin communities and migrant organizations.
7. Articulate migrant investment initiatives under a strategy oriented to build innovative regional districts, where they are already promoting social and productive projects towards a new regional productive matrix, having a higher value-added on the basis of education, science and technology. All of this, along with the implementation of new alliances with the three government levels, private sector, universities and non-governmental organizations.
8. Encouraging the constitution of a Transnational Talents Network in the following states: Zacatecas, Michoacán, Oaxaca, Puebla, Jalisco and Estado de México, as they have the main and most organized migrant presence in the US, incorporating the children, grandchildren and fellow citizens who studied in American universities to collaborate in the making of an integral and sustainable human development agenda in those regions.
9. Create in those states Virtual Training Centers for Social Economy and Transnational Community Development together with public universities and the University of California at Los Angeles and Long Beach, the private sector and the Mexican government.
10. Establish in those states a Training Program for Transnational Community Leadership focusing young migrants from the 2nd and 3rd generation. Focusing too, the 700 thousand "dreamers" who belong to the 1.2 million of Mexican migrants who entered the US as children and now face the risk of massive deportation to Mexico because Trump has reached the presidency of the US and he can overturn the temporal amnesty that President Obama had established for them some years ago.

Conclusions

The evolution of the Mexico-US migratory system, with more than hundred years, explains the formation of more than 2 thousand Mexican migrant clubs in that country and the execution of more than 26 thousand community projects by the Mexican diaspora is a proof of its philanthropy towards origin communities in Mexico. These facts show the existence of Mexican migrant organizations as a new transnational social actor that promotes local and regional social development. This actor can potentially become, with the help of other alliances, a supporter of employment and economic development in those levels and a sponsor of Solidarity and Social Economy with a transnational perspective.

Today, at the beginning of 2017, Trump's announcement of a neo-protectionism policy is coming to change radically the NAFTA with Mexico, and this would deeply affect our country's economic and employment structure. When he announces massive deportation of undocumented migrants and the possibility to set new taxes to remittances sent by migrants, it is mandatory for the Mexican government to change the corporatist and assistance perspective over migrants that limited the 3 x 1 Program and the Countrymen Program (an information program for voluntary return to Mexico). It is necessary to have an active policy for economic development and migration, able to promote regional and sector development with a strong public investment to increase national employment. Within this policy framework, it should be possible to allocate philanthropic investments as well as productive investments from the Mexican migrant organizations in the US, which are acting for more than 30 years as transnational agents for social and economic development at a local and regional level (García Zamora 2016).

In fact, since April 30, 2014, Mexico's presidency formally established the Special Migration Program after more than 2 years working with more than a hundred civil society's transnational organizations, federal government, universities, social organizations, NGO and churches. This was the effort of the America's Migration Collective, since 2013 and it represents the most advanced proposal of public policy for integral development, migration and human rights. However, until nowadays, there has been no institutional support and not enough financial support (52 million Mexican pesos in 2015 and 2016). Given the deep multidimensional Mexican crisis, and because of Trump's neo-protectionist and anti-immigration threats, there is an urgent need of financial and institutional support from the federal level and from the states with high international migration and with the highest intensity of return and migrant deportation from the US.

References

Alba Francisco. 2008. "La economía mexicana y la migración de México a Estados Unidos. Una perspectiva macro" in Agustín Escobar y Susan Martin (coordinadores) *La gestión de la migración México- Estados Unidos.: Un enfoque binacional*, Segob-Instituto Nacional de Migración. México.

Arroyo Jesús, Salvador Berumen & David Rodríguez Álvarez. 2010. "Nuevas tendencias de largo plazo de la emigración mexicana a Estados Unidos". in *Papeles de Población*. Nueva Época. Año 16, No. 63. enero-marzo.

Balboa, Juan. 2007. "En el sexenio foxista 3.4 millones de mexicanos expulsados a EU". *La Jornada*, march 4 2007. Available at: http://www.jornada.unam.mx/2007/03/04/index.php?section=politica&article=003n1pol.

El Universal. 2011. "Migración a EU cae por mejoras en México. Estudio". Redacción. *Periódico El Universal*. July 6 2011. México.

Escobar Latapí Agustín 2008. *Las políticas públicas mexicanas y la migración entre México y los Estados Unidos. Un enfoque binacional*. Secretaría de Gobernación-Instituto Nacional de Migración-CIESAS.

Fernández de Castro, Rafael, García Zamora Rodolfo y Ana Vila. 2006. *El programa 3x1 para migrantes ¿primera política trasnacional en México?* Miguel Ángel Porrúa. México.

Fernández-Vega, Carlos. 2017. "México SA". *La Jornada*. México. Available at: http://www.jornada.unam.mx/2017/01/18/opinion/022o1eco?partner=rss.

García Zamora Rodolfo 2016. "Los migrantes mexicanos como soporte de la economía y la sociedad norteamericana" in Brújula Ciudadana Number 80. Iniciativa Ciudadana. México.

García Zamora Rodolfo 2012a. *Crisis, migración y desarrollo. Los actores sociales y el reto de las nuevas políticas públicas en México*. Universidad Autónoma de Zacatecas.

García Zamora Rodolfo 2012b. "Cero Migración: Declive de la migración internacional y el reto del empleo nacional", *Revista Migraciones Internacionales*. Vol 6. No 4. July/december. Colegio de la Frontera Norte. Tijuana.

García Zamora Rodolfo 2009. *Desarrollo económico y migración internacional: Los desafíos de las políticas públicas en México*. Universidad Autónoma de Zacatecas.

García Zamora Rodolfo 2006. *Migración, remesas y desarrollo: Los retos de las organizaciones migrantes en Estados Unidos*. Universidad Autónoma de Zacatecas. Zacatecas.

García Zamora Rodolfo 2003. *Migración, remesas y desarrollo local*, Universidad Autónoma de Zacatecas. Zacatecas.

Garduño, Silvia. 2008. "Crece con Calderón la migración a EU". *Periódico Reforma*. September 24, 2008. México.

Institute for Mexicans Abroad. 2010. Paisano, invierte en tu tierra: Programa de inversión productiva para migrantes. Instituto de los Mexicanos en el Exterior/Secretaría de Relaciones Exteriores.

Milenio. 2008. "BM: México, campeón expulsor de trabajadores". Periódico Milenio. 25 de enero de 2008, Available at: http://www.dossierpolitico.com/vernoticiasanteriores.php?artid=31777&relacion=dossierpolitico.

Official Journal. 2009. "Lineamientos operativos de los proyectos productivos para el fortalecimiento patrimonial, del Programa 3×1 para Migrantes". Diario Oficial de la Federación, 29 enero 2009. (Published in DOF with code 5079086)

Orozco Manuel. 2004. "HometownAssocations and TheirPresent and Past-Partnership: New DevelopmentOpportunities" Report commissioned by the U.S. Agency for International Development. InterAmerican Dialogue. Available at: http://meme.phpwebhosting.com/~migracion/modules/documentos/manuel_orozcohome.pdf.

Presidencia de la Republica. 2016. Cuarto Informe Presidencial. Available at: http://www.presidencia.gob.mx/cuartoinforme/.

Tuirán Rodolfo, Rodolfo Corona y José Luis Ávila. 2000. "Las causas de la migración hacia los Estados Unidos" in Rodolfo Tuirán (ed) *Migración México-Estados Unidos. Presente y futuro*, Consejo Nacional de Población, México.

Ethiopia and Role of Its Diaspora in Development

Rashmi Mourya

Introduction

Migration is an inherent, continuous, and spatial process through which individuals relocate from one region to another. This movement contributes to the formation of diasporic communities. According to Sheffer (1986: 3), "modern diasporas are ethnic groups of migrant origin residing and acting in host countries but maintaining strong sentimental and material links with their countries of origin- their homelands". Diasporic communities, though residing beyond their countries of origin, continue to maintain deep-rooted connections with their homelands and are expanding in both scope and influence. These communities play a pivotal role in shaping domestic political dynamics and exerting influence on international relations and foreign policy.

Migration has been an essential part of Ethiopian History. In the last forty years, Ethiopia has been experiencing socio-economic, environmental, and political crises that have led to a massive migration of people, both internally and cross-borders. These internal political, economic, and social crisis are reasons for the cross-border flow of Ethiopian migration

R. Mourya (✉)
Govt. Lochan Prasad Pandey PG College, Sarangarh (C.G), Shaheed Nandkumar Patel University, Raigarh, Chhattisgarh, India

A. Dubey and S. Rajwar (eds.), *Diaspora in Development*,
https://doi.org/10.1007/978-981-96-6747-5_11

and become a major factor in the formation of the Ethiopian diaspora (Gulema et al. 2024).

Globalisation has created diverse avenues that enable diasporic engagement across cultural, economic, and political domains. In the era of intensified globalisation, the nexus between diasporas and the socio-economic development of their home countries has become increasingly significant (Condé 1998). Diasporas contribute to their countries of origin not only through direct financial investments and remittances but also by acting as intermediaries in various economic and social domains. They facilitate trade by connecting businesses from their host and home countries, leveraging their knowledge, networks, and trust built in both regions (Kapur 2010). They play a key role in introducing and implementing new technologies. By transferring technical knowledge, innovative business practices, and expertise gained in more advanced economies, they contribute to the modernisation and development of industries in their home countries (Kapur 2001). The economic impact of diaspora on their home countries is primarily studied through their significant financial contributions, particularly in the form of remittances. In many low-income countries, remittances are the largest source of net foreign exchange inflows, surpassing capital market investments, FDI, and public development aid. This consistent flow is largely due to the continuous nature of migration, as diaspora communities maintain financial ties with their families (Ratha et al. 2011). This chapter will use the case of Ethiopian diaspora to examine the crucial role in the country's development through economic contributions, political influence, and transnational networks. It also assess the diaspora's impact that extends beyond monetary transfers and includes knowledge exchange, investment, advocacy, and diplomatic engagement in overall development of Ethiopia. The data were analysed using data from the World Bank, IOM, and UNCTAD.

History of Ethiopian Migration

According to estimation over 2.5 million Ethiopians are living outside of the country. This is the result of a series of migrations which has been taking place throughout history. Historically, Ethiopians have migrated for trade, religious, scholarship, and diplomatic purposes. Post-World War II, Ethiopian students and intellectuals travelled abroad for education, particularly to Europe and North America (Zewde 2002). There are four major waves of migration: before the 1974 revolution, during the Qey shibir (1974–1982), family reunification between 1982 and 1991, and the post-Dragument period (1991 to present), where people fled from ethnic violence and political repression. These waves have impacted Ethiopia's economy, education, and social structure. The liberalisation of the Ethiopian economy in the 1990s led to increased emigration as individuals sought better economic opportunities abroad. Ethiopian migration has continued into the twenty-first century, driven by factors such as political instability, ethnic tensions, economic hardship, and environmental challenges.

Ethiopians' large emigration from Ethiopia has happened from 1970s when Ethiopia faced political persecution and government transition. Ethiopia, like many other African military administrations, experienced nearly two decades of cruel Derg tyranny, assumed power in 1974 then the peasant farmers, and political activists left Ethiopia for various places, including Kenya, the United States, and Sudan while others travelled further to the Middle East, Europe, and North America. During the late 1970s and early 1980s, approximately 15,000 Beta Israel, also known as Ethiopian Jews, made the journey from the northern highlands to Sudan and later were transported by air to Israel. Subsequently, in the early 1990s, numerous additional Beta Israel individuals also migrated to Israel. During the military dictatorship, Ethiopia saw the worst famine, human rights violations, political turmoil, and economic hardship in recorded history, as well as an unprecedented level of emigration. Many of these migrants were relocated to Western nations, including the United States, Europe, and Australia, from adjacent refugee camps. Between 1982 and 1990, Ethiopian refugees were the majority (90%) of African refugees in the United States (Chacko and Gebre 2013).

The military dictatorship collapsed in May 1991, it was succeeded by the EPRDF. The nature of human migration has shifted significantly since 1991 as a result of major structural changes in governmental and development policies, as well as political turbulence and war with neighbouring nations. Over the past two decades, Ethiopian migration patterns have shifted due to rising unemployment and poverty. Female domestic workers are moving towards the Middle East and Gulf states, particularly Saudi Arabia, for better employment and living conditions. Traditionally associated with northward migration, Ethiopia also experiences significant flows towards southern African countries like Kenya, South Sudan, Sudan, Uganda, and South Africa. The Horn of Africa is the primary source for irregular migration from Ethiopia, Somalia, and Kenya to South Africa, with two-thirds of smuggled migrants originating from Ethiopia.

The migration of domestic workers from Ethiopia to the Gulf countries and the Middle East has been significant over the past few decades, driven by economic factors, demand for domestic labour, and government policies. In recent years, Lebanon, Saudi Arabia, and Kuwait have become popular destinations for Ethiopian domestic workers with limited access to education and employment. According to data from the Ministry of Labour and Social Affairs (MoLSA), the number of documented migrants departing for Arab countries increased from 30,000 in 2009 to 200,000 in 2012, with women accounting for approximately 95% of all documented migration out of Ethiopia, with an average of 805 women registered each day (Ministry of Labour and Social Affairs 2010). This labour migration is both voluntary and forced, with migrants using both channels. Yemen, the closest Arab nation to the Horn region, remains a major destination for desperate migrants from Ethiopia.

Mapping of Ethiopian Diaspora: A Global Overview

According to the International Organisation for Migration, around 2.5 million Ethiopians live outside the country as part of a diaspora. The most popular destinations being the Middle East, North America, Europe, and Asia. Major Middle Eastern destinations include Israel, Saudi Arabia, Yemen, and Lebanon, while African destinations include Somalia, South Sudan, Eritrea, Sudan, and South Africa. Kenya and Djibouti are also popular destinations. The United States, Saudi Arabia, and Israel are the top three destinations for Ethiopian migrants in individual countries (Table 1).

According to UNDESA (2017), the United States is the primary destination for Ethiopian emigrants. Ethiopian immigrants have traditionally arrived in large numbers in the United States. Both highlighted four stages of Ethiopian Diaspora inflow. The first happened before 1974 and comprised mostly of elites. The second wave arose between 1974 and 1982 when people fled the Derg's Red Terror. The third wave, which persisted from 1982 to 1991, was mainly focused on family reunions in the West. The fourth wave started after 1991, when people fled to escape ethnic conflict and political persecution.

Table 1 Ethiopian migrants by country of destination 2000–2015

S. N	*Destination*	*2000*	*2005*	*2010*	*2015*
1	USA	71,296	109,858	152,798	210,917
2	Saudi Arabia	64,252	79,371	102,910	131,493
3	Israel	57,163	67,189	78,029	80,474
4	Sudan	123,478	107,732	70,677	60,734
5	Kenya	20,846	20,917	24,091	36,889
6	Italy	5569	17,983	30,396	30,401
7	Canada	14,208	18,889	24,535	27,438
8	Germany	14,795	15,936	16,835	17,541
9	Sweden	11,907	11,221	13,100	16,246

Source United nations department of economic and social affairs (2017)

Before 1974, Emperor Haile Selassie actively encouraged foreign involvement and placed a strong priority on improving Ethiopian education. He sent top-achieving students abroad, mostly to North America and Europe, with a preference for the United States in the 1960s and 1970s. High-achieving Ethiopian students were sent abroad annually to study in Europe and North America between 1930 and 1974 (Zewde 2002). However, a new wave of immigration that encompassed larger and more diverse groups of people was brought about by the 1974 Revolution in Ethiopia. The number of Ethiopians leaving the country in search of safety and security increased as a result of these driving forces. A Migration Policy Institute (2014) reports that 60% of Ethiopian immigrants to the US came here after 2000. From 35,000 in 1990 to over 220,000 in 2017, Ethiopians have emigrated to the US at a sharply higher rate. The United States has seen an increase in the number of migrants, including Ethiopians, as a result of the 1965 Immigration Act, the Refugee Act of 1980, and the Immigration Act of 1990's Diversity Visa Program. The majority of Ethiopian immigrants came in the 1990s, following the Ethio-Eritrean War and a change of government. With over 251,000 Ethiopian immigrants and their children residing in the US as of 2014, they are the second-largest group of African immigrants after Nigeria (MPI 2014).

After the US, Canada is the second-largest North American destination for Ethiopian emigrants. The changes to Canada's immigration policies in the 1970s helped the country's substantial Ethiopian refugee population. The first Ethiopian refugee groups were admitted to Canada under the Immigration Act of 1976, which also created new immigration categories. Ethiopians started coming to Canada by applying for asylum or being admitted as refugees while they were elsewhere. According to the Toronto Ethiopian Association, there are at least 50,000 Ethiopians living in the Greater Toronto area alone. Between 1976 and 1993, a large number of Ethiopian immigrants settled in Canada. In Canada, there are around 44,065 Ethiopian immigrants, according to the 2016 Census. About 15,990 people of Ethiopian descent lived in Toronto in 2016, with Calgary (6355 immigrants), Edmonton (5210 immigrants), and Ottawa (2850 immigrants) following closely behind (Statistics Canada 2017).

Ethiopians have also relocated to Europe for a variety of reasons, including improved economic prospects, education, protection from political persecution, and family reunification. According to the European Commission, the top destinations for Ethiopian migrants in the EU in 2016 were Sweden, Germany, Netherland, and Italy. By 2009, there were just 9990 Ethiopians living in Germany. These number of Ethiopians grew significantly in 2015. The formal Ethiopian diaspora in Germany is estimated to be about 24,500 people, including 8000 Germans of Ethiopian descent and about 2,000 second-generation Ethiopians who are German citizens (Schlenzka 2009). The Ethiopian diaspora in the Netherlands has gradually increased throughout the years. Following asylum, education is the primary motivation for Ethiopian migration to the Netherlands. A quarter of Ethiopian students study in the Netherlands, mostly at the Institute for Social Studies in Den Haag and the Agricultural University of Wageningen.

Ethiopian immigrants to Israel, another prominent settlement destination, are mostly Ethiopian Jews, commonly known as 'Beta Israel' (BenEzer 2002). In the 1950s, a small group of Ethiopian Jews began migrating to Israel, using the Law of Return to reunite their scattered Jewish community. However, in 1974, the Derg prohibited emigration and closed Ethiopia's borders, leading to persecution and restrictions on religious communities. Thousands of Ethiopian Jews made the challenging journey to Israel via Sudan and Kenya, with up to 4000 believed to have died along the way (BenEzer 2002). According to Israel's Central Bureau of Statistics (2017), approximately 155,300 Ethiopian Israelis were residing in the nation, accounting for 1.7% of the total population. Approximately half of the population was born in Ethiopia, with the remaining born in Israel to an Ethiopian-born father. The majority of Ethiopian immigrants arrived in two waves: Operation Moses in the mid-1980s and Operation Solomon in the early 1990s (BenEzer 2002).

Saudi Arabia is a popular destination for Ethiopian migrants. Saudi Arabia's diverse industries include construction, hospitality, domestic labour, and agriculture, offering job opportunities for Ethiopians seeking better opportunities. According to Ministry of Labour and Social Affairs (MoLSA) data from 2008–2009 to 2012–2013, three regions (Oromia, Amhara, and Addis Ababa) alone contributed 316,064 legal migrants to the Gulf. According to an Ethiopian MoLSA (2010) report, between July 2008 and July 2009, the agency supplied 21,256 job contracts licensed by PEAs (private employment agencies) to Ethiopians working in the Middle East. Eighty-two percent of these work permits were for women to work in Saudi Arabia (61.9%). Religion and religious pilgrimages have also contributed significantly to the continued migration from Ethiopia to Saudi Arabia. The primary incentive for travelling is to escape poverty, improve family living conditions through remittances, and protect family businesses. Ethiopians have migrated all over the world, yet they have never lost their sense of identity or commitment to their homeland. The Ethiopian diaspora formation is deeply rooted in the historical identity of Ethiopianism, which emphasises togetherness and pride in Ethiopian culture and history, is closely related to the formation of the Ethiopian diaspora.

Ethiopian Diaspora and Foreign Policy Implication

Joseph Frankel (1968) defines foreign policy as "consists of decisions and actions which involve to some substantial extent relations between one state and others". Nowadays, the diaspora has become an intrinsic component of national states' foreign policy discourse. The formulation of a country's foreign policy is influenced by a range of internal and external factors. In contemporary times, diasporas have emerged as significant actors in redefining conventional understandings of borders and citizenship. This transformation has contributed to the development of a "politics of belonging" (Shain 2007). Acknowledging the growing influence of diaspora communities, many states are increasingly revising or adjusting their foreign policy frameworks to strengthen engagement with their diaspora and harness their potential for national development. Diaspora are seen as a soft power in foreign policy strategy, as well as an agent or catalyst for economic growth in their home countries, in addition to their active engagement in host nations (Khara 2020). Indeed, the contributions of the diaspora are not confined to financial flows only, but also to knowledge transfer, entrepreneurship, and serving as important agents of trade between the country of origin and country of destination.

In Ethiopia, the transition from military rule led to a significant shift in the country's foreign policy approach and orientation. This transformation culminated in the adoption of the Foreign Policy and National Security Strategy in 1996, marking the first formal articulation of Ethiopia's national interests and security threats. The policy framework identified internal vulnerabilities particularly economic underdevelopment and poverty as primary threats to national security. It emphasised that the persistence of widespread poverty, coupled with the necessity for democratic governance and institutional reforms, posed existential risks to the nation. During the EPRDF-era, Ethiopia's foreign policy main objective was securing political autonomy to support economic development and transformation. This goal was explicitly rewritten in Ethiopia's 2002 "Foreign Affairs and National Security Policy and Strategy" White Paper (Ministry of Information 2020). Ethiopia's foreign policy was largely shaped by a developmental state model, emphasising economic diplomacy, regional stability, national interests, regional dynamics and strategic partnerships.

In the wake of globalisation changes in the Ethiopian economy, Ethiopian Diaspora is considered a viable and potential source of foreign currency and remittances. The engagement with the Ethiopian Diaspora resulted in major implications for internal as well as external political and economic processes for Ethiopia. The substantial presence of a diverse Ethiopian community in various parts of the world has also had implications on Ethiopia's diplomatic relationship with many countries. The Ethiopian diaspora's role in shaping Ethiopian foreign policy is significant and multifaceted, encompassing economic, cultural, and political dimensions.

Ethiopian Government Diaspora Engagement Efforts

Since 2002, the Ethiopian government's policy framework has increasingly emphasised the role of the Ethiopian diaspora in the country's development. This strategic focus aligns with the broader shift towards recognising diasporas as key contributors within the migration-development nexus. The government of Ethiopia established formal mechanisms and enacted several laws and legislations to promote and enable diaspora engagement within the nation to maximise the utilisation of diaspora resources.

The Ethiopian government has established several institutions to strengthen its relationship with its diaspora. The Ethiopian government established a General Directorate of Ethiopian Expatriate Affairs inside the Ministry of Foreign Affairs in 2002 as part of a program to incorporate the diaspora into the nation's socio-economic growth (Chacko and Gebre 2013; Kuschminder and Siegel 2013: 58). At present time, MoFA Ethiopia plays a pivotal role in diaspora engagement, serving as a key facilitator and coordinator of initiatives aimed at fostering strong ties with the Ethiopian diaspora community worldwide. Through diplomatic missions and outreach programs, the MoFA aims to strengthen relationships with Ethiopian communities abroad and foster a sense of belonging and partnership that contributes positively to Ethiopia's growth and prosperity.

The Employment Service Protection Directorate of the MoLSA plays a key role in the emigration of Ethiopian migrant workers. This section is in charge of creating policies to defend migrant workers' rights, promoting legal migration, protecting Ethiopian migrant workers overseas. The EIA was created by the Ethiopian government to manage all foreign investments, provide investors with pre- and post-investment services, and communicate with other pertinent organisations. The Ministry of Trade and Industry oversees the EIA. The National Bank of Ethiopia also regulates finances and remittance-related affairs. The NBE is in charge of Ethiopia's Development Bank, which provides finance for regional development projects. The DBE facilitates the diaspora's access to credit. The DBE has put policies in place to enable diaspora persons to acquire loans more rapidly (Berhanu 2019: 14).

The government's development goals, which recognised the diaspora as a source of funding to close the nation's external resource deficits, reiterated the need for the government to establish an atmosphere that is conducive to diaspora involvement. For example, in its 2002–2003 "Sustainable Development and Poverty Reduction Program" (SDPRP) paper, the government suggested that policy initiatives that included investments by Ethiopians in the diaspora would increase the private sector's share of the economy. The government set lofty targets to eradicate poverty and promote long-term development in the years that followed, with the diaspora mostly seen as a source of funding.

Over the time Ethiopian government introduced several proclamation to moblise Diaspora resource. Among such initiatives the proclamation 270/2002 provides right, anyone who has even obtained citizenship in another country but whose origin is Ethiopia has the right to obtain a special identification card that grants them exclusive rights as foreign nationals of Ethiopian descent. The ID card holder is entitled to certain benefits and privileges, such as the ability to enter the nation without a visa, live there without a residency permit, work in Ethiopia without a work permit, be treated as a domestic investor, and make investments in areas that are open to Ethiopian investors without facing any discrimination. One another initiative has taken to integrate diaspora in development was the Investment Proclamation No. 280/2002, and its subsequent amendments, including Proclamation No. 1180/2020, provide members of the Ethiopian diaspora the legal status to be treated as domestic investors in Ethiopia. The Investment Proclamation No. 847/2014 allows diaspora investors to fully own businesses in various sectors without the requirement for local partnerships, thereby enhancing their participation in the local economy.

The NBE Directive No. FXD/30/2006 and its 2009 revision are the most important of these measures. The "Provisions for International Remittance Services" improve incoming remittance transfers, including lowering remittance prices and expanding access to dependable, quick, and secure remittance services in Ethiopia. Another directive adopted by FDRE in 2006 was Directive No. FXD/31/2006, permitting Ethiopians who are not residents to open foreign currency accounts.

In 2008 Ethiopia was the second country in sub-Saharan Africa to issue a diaspora bond. The diaspora Bond was issued by the state-owned power utility company Ethiopian Electric Power Corporation (EEPCO). The Corporate bond known as the Millennium Bond is underwritten by the Commercial Bank of Ethiopia. Additionally, the NBE has worked with Ethiopia Electric and Power Corporation to issue diaspora bonds to encourage the diaspora to contribute to significant energy-producing projects. The Grand Ethiopian Renaissance Dam (GERD) bond issue was specifically targeted to the Ethiopian diaspora population in 2011 as part of a broader effort by the Ethiopian government to generate funds domestically for the dam's construction. GERD bonds were offered to encourage Ethiopians living abroad to contribute financially to this significant infrastructural project. The government's belief that the Ethiopian diaspora can contribute significantly to the country's growth through knowledge and technology transfer, investment, and remittances. In 2013, the Ethiopian government promulgated the Ethiopian Diaspora Policy to establish a comprehensive legal framework for regulating its interactions with the Ethiopian diaspora. The primary objective of the policy is to formalise and institutionalise mechanisms for engagement with the diaspora community, in recognition of the increasing significance of Ethiopians living and working abroad, particularly in the domains of remittances, investment, knowledge transfer, and political influence (Ministry of Foreign Affair 2013). The establishment of the Ethiopian Diaspora Service under Regulation 432/2019 marks a significant milestone, aimed at implementing the diaspora policy and safeguarding diaspora rights. The New Investment Proclamation No. 1180/2020, New Investment Regulation No. 474/2020, and Investment Incentive Regulation No. 270/2012 have expanded opportunities for the Ethiopian diaspora to engage in 28 sectors. Under Ethiopia's investment laws, members of the diaspora have the right to choose whether to be recognised as local investors. As a result, the Ethiopian diaspora is now eligible to invest in multiple key sectors contributing to the country's development.

Ethiopian Diaspora Role in Development

The Ethiopian diaspora plays a pivotal role in the socio-economic landscape of Ethiopia, serving as a valuable resource for national development (Kuschminder and Siegal 2013). The government of Ethiopia has recognised the importance of engaging its diaspora community to harness its skills, expertise, and financial capital for economic growth. At present in Ethiopia, diaspora investments and contributions have influenced key sectors of the Ethiopian economy, such as agriculture, manufacturing, and infrastructure. A key indicator of this evolving perspective is the substantial remittances sent by Ethiopian nationals abroad, reflecting their economic engagement.

Economic Dimension

The Ethiopian government recognises the diaspora's potential to contribute to Ethiopia's growth and has established various initiatives to facilitate diaspora investment and treats them as domestic investors. In 2002, the Investment Proclamation No. 280/2002 expanded the definition of a Domestic Investor to include foreign nationals of Ethiopian descent. As a result, diaspora members are treated as domestic investors and enjoy more advantageous rights compared to foreign investors in Ethiopia. The Council of Ministers Regulations No.84/2003, passed in 2003, outlines the rights of domestic investors and offers incentives such as import duty exemptions and income tax breaks. It also designates certain industries exclusively for domestic investment amendments have made investing in Ethiopia more accessible for diaspora members, with approximately 10% of domestic investment in the last decade coming from the diaspora, predominantly in Addis Ababa (Kuschminder and Siegal 2013).

The EIC offers incentives such as tax holidays, duty-free imports of capital goods, and land lease opportunities to diaspora investors. For example, under the Investment proclamation No. 769/2012, diaspora investors can benefit from incentives like income tax exemptions for up to five years and customs duty exemptions on imported capital goods. The Investment Proclamation and other related laws ensure that diaspora investors have secure ownership rights and can repatriate profits and dividends earned from their investments. Council of Ministers Regulation No. 270/2012 in Ethiopia outlines investment incentives and designates specific investment areas exclusively for domestic investors. This regulation benefits the Ethiopian diaspora by offering opportunities to participate in reserved investment sectors and access various incentives aimed at promoting local economic development. Ethiopian diaspora members are eligible to invest in sectors designated for domestic investors under this regulation. By investing in these reserved areas, diaspora investors are contributing to the growth of key industries such as agriculture, manufacturing, and services, which are vital for Ethiopia's economic advancement.

The governments of developing nations are formulating and implementing policies to leverage the resources of their diaspora for national development (Ratha et al. 2011). Ethiopian Diaspora members play an important role in the homeland economy by sending back remittances to Ethiopia. In recent years, the Ethiopian government has implemented several initiatives to enhance formal remittance channels, lower transfer costs, and expand access to international remittance services. In 2011, Ethiopia became a signatory to the Maya Declaration, pledging to modernise its national payment systems, enhance financial accessibility, expand digital financial services, and promote financial literacy.

According to the Ethiopian Investment Report (2019), remittances from the Ethiopian diaspora have become a major source of external funding, surpassing Overseas Development Assistance, Foreign Direct Investment (FDI), and export revenues. Studies indicate that approximately 60% of remittance inflows are allocated toward household consumption, healthcare, and education. According to World Bank (2024) data, Remittance has averaged 1.36% of GDP over the last 16 years. Between 2002 and 2023 remittance flows have steadily grown from 33 million USD (in 2002) to 1.8 billion USD (in 2014). These inflow remittances are providing finance in the country for investment which is promoting economic growth. The country's economy is greatly strengthened by this inflow of money, which supports many industries and promotes general expansion. The diaspora remittances have been steadily increasing since 2010, with over $436.1 million being sent home by Ethiopian diaspora members. This amount has increased dramatically to almost $624.3 million by 2012, which accounted for a hefty 1.4% of Ethiopia's GDP. Ethiopia witnessed a remarkable increase in remittance inflows in 2014. Despite the ongoing global economic crisis, the country received a staggering $1.8 billion in remittances during that year, approximately 3.2% of the country's GDP (see Fig. 2). In addition to remittances, Ethiopia's Investment Agency documented a noteworthy pattern of Ethiopian diasporas between 2003 and 2013. They are actively involved in investment projects; they have started 1,906 projects in Ethiopia and raised an outstanding $1.92 billion in investment money (EIA 2013) (Fig. 1).

In the early 2000s, Ethiopia leveraged diaspora bonds to finance power plant projects. To fund the construction of the Grand Ethiopian Renaissance Dam (GERD), a major hydroelectric facility, the government issued bonds to its citizens, the diaspora, and other investors. This strategy proved successful, generating $56 million since GERD's inception. Notably, in 2014, the Ethiopian diaspora purchased diaspora bonds worth $22 million. In recent years, the Ethiopian government has actively encouraged diaspora engagement and investment. As a result of

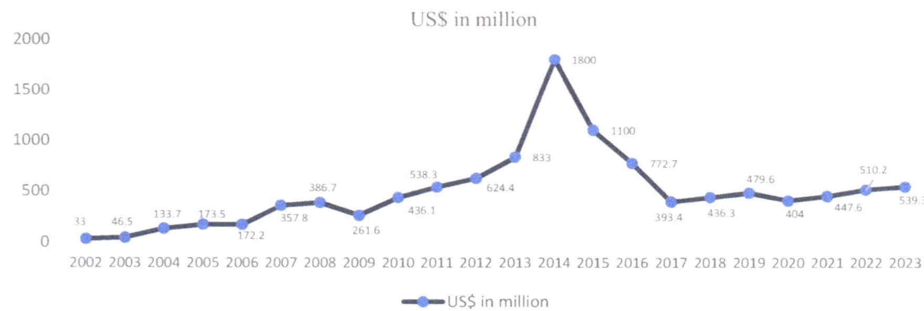

Fig. 1 Remittance flow in Ethiopia, 2002–2023 (million in US $). *Source* World Bank (2024)

Fig. 2 Personal Remittance received in Ethiopia, 2002–2023 (% in GDP). *Source* World Bank (2024)

these initiatives, between 1992 and mid-2009, the Ethiopian Investment Agency (EIA) issued over 1,800 investment licenses to Ethiopians living overseas, with 39% of recipients being residents or citizens of the United States (Chacko and Gebre 2013). Most of these investments focused on real estate development and manufacturing, while notable contributions were also made in sectors like hospitality, construction, healthcare, education, and agriculture (Chacko and Gebre 2013). Moreover, the Ethiopian Ministry of Foreign Affairs (MoFA) reported that in the first ten months of 2016, remittance inflows surpassed the country's earnings from exports. At the household level, remittances represent a critical source of income for many families. Given their large volume, these financial inflows have the potential to be redirected toward productive investments, fostering Ethiopia's long-term economic development.

Recognising the diaspora's role in FDI is essential, as it is widely acknowledged as a key factor in driving economic growth and national development (Solomon 2008). To enhance the overall investment climate and solve issues with attracting FDI into different sectors, the Ethiopian Federal Government has implemented several regulatory frameworks since 1992. These initiatives are in line with a market-oriented economic strategy. To expand sectoral coverage and satisfy the demands of both domestic and international investors, the investment legislation has undergone several revisions. The initial Investment Proclamation No. 15/1992, was enacted in 1992. The current regulatory framework comprises Investment Proclamation No. 280/2002, Investment (Amendment) Proclamation No. 84/2003, and Investment (Amendment) Proclamation No. 146/2008 to encourage FDI.

The FDI data (Fig. 3) illustrates the FDI trend from 2002 to 2022, as reported by UNCTAD. The FDI flows remained relatively low and fluctuated modestly during this period. From 2002 to 2004, FDI gradually increased from $255 million to $545 million, indicating initial efforts to attract foreign investment. However, between 2005 and 2010, there were significant fluctuations, with FDI dropping to a low of $109 million in 2008, likely due to global financial crises and domestic constraints. A notable upward trend began after 2010. By 2013, more than 3,000 diaspora members had invested over $1 billion, mostly in the growing real estate and hotel and tourist industries, with 91% of this investment focused in Addis Ababa (EIA 2013). FDI peaked in 2021 at $4259 million, reflecting investor confidence and Ethiopia's improved business climate during this period. The overall FDI trend demonstrates substantial growth from 2013, with FDI increasing more than tenfold from $255 million in 2002 to $3670 million in 2022. This growth underscores Ethiopia's success in positioning itself as an attractive investment destination, particularly in sectors like agriculture, manufacturing, and infrastructure.

The Africa CEO Survey ranked Ethiopia as the sixth-most attractive African country for investment in 2020. There are different sectors of the Ethiopian economy where diaspora members are playing crucial roles. Many Ethiopian diaspora entrepreneurs have established textile factories in industrial parks like Hawassa Industrial Park. In the Somali Region, agriculture and industry have emerged as key investment sectors over the past five years. The regional administration has acknowledged these areas

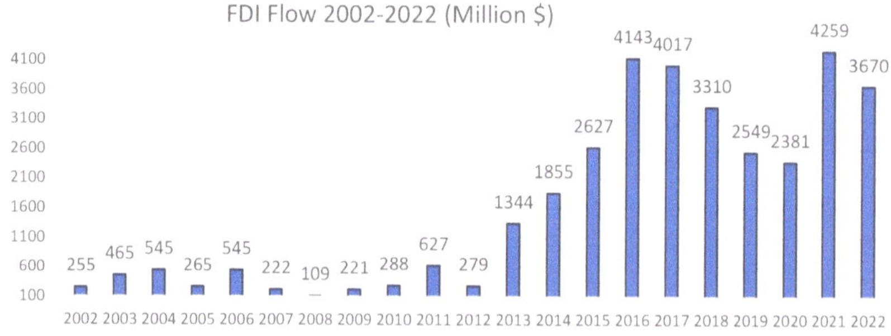

Fig. 3 Ethiopia Foreign Direct Investment from 2002–2022 (million in US $). *Source* UNCTAD (2023)

as significant contributors, where diaspora investors are playing a crucial role in generating employment opportunities and facilitating skill transfer to the local community.

Diaspora engagement in trade often involves promoting Ethiopian products and brands in international markets (Getahun 2019). Diaspora entrepreneurs are vital in promoting trade between Ethiopia and the world, particularly in the coffee industry. Ethiopian coffee, known for its quality and cultural heritage, is favoured globally. Diaspora members, particularly in the US and Europe, promote and distribute Ethiopian coffee to international markets. Debritu Mogese, an Ethiopian diaspora member, has invested in Moyee Coffee, exporting premium coffee globally. She is also mobilising the Ethiopian diaspora community through Connecting Diaspora for Development initiatives through IOM Netherlands. Ethiopian diaspora investments in construction materials and building supplies contribute significantly to infrastructure development and urbanisation in Ethiopia. A notable example in this area is Derba MIDROC Cement, which is owned by MIDROC Ethiopia, a company controlled by Sheikh Mohammed Hussein Al-Amoudi, an Ethiopian-born Saudi billionaire and diaspora investor. He also owned Saudi Star Agricultural Development Plc plans to develop up to 500,000 hectares of Ethiopian land for sugar and grain production.

Prior to the enactment of the diaspora policy in 2013, investment from the diaspora in the Somali region was relatively low. One significant area of business engagement among Ethiopian Somali diaspora members is trade, particularly in the export of processed meat. For instance, the Jigjiga Export Slaughter House Plc (JESH), founded in 2013 with an initial investment of 200 million Birr by a returning member of the Ethiopian Somali diaspora, exports processed meat to the United Arab Emirates (Berhanu 2019). According to 2016 data from the Ethiopian Somali Regional State Ministry of Diaspora, the hotel industry attracts the highest number of diaspora investment projects, making it the most prominent sector. Construction ranks third in popularity, while real estate holds the sixth position. A large share of these investments is concentrated in Jigjiga town.

Ethiopia's agricultural sector holds vast potential for enhancing productivity and increasing rural incomes (Shepherd 2013). Data from the Benishangul Regional Investment Bureau indicate that out of 227 agricultural investors, 40 approximately 18% belonged to the diaspora. Additionally, the Agricultural Investment Support Directorate under the

Ethiopian Ministry of Agriculture reported that, at the national level, 24 land lease agreements were signed between 2009 and 2011, with six of these allocated to Ethiopian diaspora members. Collectively, these leases cover 350,099 hectares of land, of which 65,087 hectares are held by Ethiopians, including diaspora investors (Bossio et al. 2012). Based on the available data, both Ethiopian and diaspora investors primarily secure land for cultivating cotton, grains, horticultural products, and oil crops. Investments in the agricultural sector are anticipated to enhance food security in the near future (Shepherd 2013).

The Ethiopian diaspora is a significant player in the country's real estate market, supporting investment inflows and developmental initiatives. Successful examples include condominium development projects in Addis Ababa, where diaspora individuals or groups invest in residential properties. Following the 2008 financial crisis, members of the Ethiopian diaspora increasingly turned to property development in Addis Ababa. Goodfellow (2017) notes that a real estate developer reported that 75% of their buyers were Ethiopians residing abroad, primarily in the United States. Between 1994 and 2014, 60% of all investment by the diaspora was in real estate and related services, four times more than in manufacturing, 91% of all diaspora investment was located in Addis Ababa (Goodfellow 2017).

Diaspora business owners often invest in education and training programs in Ethiopia. They may support schools, vocational training centres, or specific skill development initiatives to enhance the local workforce's capabilities. A diaspora-owned company STEM Synergy is funding scholarships for Ethiopian students to pursue STEM (Science, Technology, Engineering, and Mathematics) education in Ethiopia. Diaspora entrepreneurs have established technology hubs and startup incubators in Ethiopia, such as "caddis" and "blue-moon". These platforms provide training and mentorship for local innovators, promoting technological skills. Diaspora business owners support entrepreneurship by establishing incubation centres and providing seed funding for start-ups. These initiatives encourage business skills and promote innovation in various sectors of the economy.

Overseas Ethiopian diaspora organisations play a crucial role in promoting economic engagement and development in Ethiopia through various initiatives and collaborations. One example is the Ethiopian Diaspora Trust Fund (EDTF), established by the Ethiopian government in partnership with the diaspora community. EDTF mobilises financial contributions from overseas Ethiopians and allocates funds to projects in Ethiopia aimed at improving education, healthcare, and infrastructure. As of 2021, the EDTF raised over $5 million from over 25,000 global contributors, funding clean water access in Amhara and Oromia regions, school construction, and rural healthcare initiatives (EDTF 2024). EDTF has allocated $500,000 in medical aid to address the urgent shortage of essential medicines and medical supplies in Ethiopia. In recent years, the fund has successfully mobilised over 26,000 donors across 93 countries, established 48 chapters, and enlisted the support of more than 200 volunteers. Additionally, diaspora organisations like the Ethiopian American Chamber of Commerce (EACC) facilitate trade and investment by connecting businesses in Ethiopia with markets and opportunities abroad. Through these and similar initiatives, overseas Ethiopian diaspora organisations foster economic engagement, knowledge transfer, and sustainable development in Ethiopia.

Diaspora organisations like the Ethiopian Diaspora Association in Germany (EDAG) facilitate investment platforms that connect Ethiopian diaspora investors with promising business opportunities in Ethiopia. For instance, EDAG organises investment forums and matchmaking events where entrepreneurs and investors can network and explore investment possibilities (Mekonnen and Lohnert 2018). The Ethiopian Diaspora Community Association (EDCA) funds initiatives such as road construction, school renovations, and healthcare facilities in rural areas. These projects create employment opportunities and improve livelihoods. All these examples indicate that diaspora associations not only doing philanthropic activity but also, they are encouraging economic engagement within Ethiopia.

Strategic Dimension

Kapur (2003) stated that a diaspora can be a strategic resource to a country when acting as advocacy groups to advance the interests of the home country; promotion of cultural richness and bringing human capital with high knowledge potential; brain circulation through innovation, investment, and expansion of businesses; assistance during emergencies; relief work; more active involvement in reconstruction and peacebuilding through diaspora associations (Kapur 2003). Ethiopian diaspora abroad play a dynamic role in connecting Ethiopia to the host countries. They promote the interests of Ethiopians living abroad and facilitate opportunities for investment, lobbying efforts, and policies that work according to the needs of the community. There are so many development-oriented organisations within the community, comprised of highly educated professionals with expertise in areas such as education, health, and engineering. For instance, The AHEAD Ottawa, ENAHPA, Sahay Solar (Germany) is a prime example, as it concentrates on environmental, agricultural, and energy projects in Ethiopia and Tigray Development Association (TDA) (Mekonnen and Lohnert 2018: 9). The Tigrayan diaspora established the Tigray Development Association (TDA) in Washington, D.C. in August 1989. TDA was founded with the goal of assisting Tigrayans residing abroad and linking them to their home country, as well as advocating for the rehabilitation of the Tigray people displaced by the 1970s and 1980s violence and hunger. It has gradually grown into a large development organisation, now operating in Tigray, Ethiopia. Sahay Solarverein Africa (SSA), formed in 2009 by Ethiopian and German Diaspora, aims to promote off-grid solar energy in rural Ethiopia. Their goal is to establish a competence centre for solar technology at Arba Minch University (AMU) in southern Ethiopia, enabling long-term dissemination of this renewable energy source (Mekonnen and Lohnert 2018: 11). Currently, SSA is working to provide solar energy to 50 rural Health Centres in southern Ethiopia's Gamo Gofa Zone, showcasing the collaboration between private, university, and diaspora initiatives.

The diaspora, including Ethiopians living abroad, possess specialised knowledge in fields like engineering, medicine, sciences, arts, and business. Their expertise can be used to improve institutions, research projects, and capacity-building efforts in host nations. This, in turn, benefits Ethiopia by bringing a range of skills and expertise that contribute to its development. Their impressive academic accomplishments help in collaborative research and shared publications inspire academics to participate in research endeavours and exchange findings. This brain drain of professionals from higher education institutions hampers various aspects of Ethiopia's progress particularly within the healthcare sector. From 1987 to 2006 Ethiopia lost 3704 medical professionals (Degu 2021). To revert brain drain, existing offices playing significant roles, such as the Alliance for Brain Gain and Innovative Development (ABIDE) and the Diaspora Affairs Directorate at the MoFA are now working great in trapping synergistic partnerships with intellectual residing outside of the country. The intellectual Ethiopian diaspora, like the Ethiopian North

American Health Professionals Association, has positively impacted their home country by providing free medical services (Degu 2021). Ethiopian government incorporates health professionals of the diaspora through numerous ways like the Diaspora Health Program and Diaspora Fellowship Program. Ethiopian diaspora Association also working in the health sector, for instance, the Tigray Diaspora Association (TDA) has extended its involvement in the health sector in Tigray. TDA made its initial contribution by providing an ambulance, 20 containers filled with medical equipment such as hospital beds, x-ray machines, and dialysis machines the first of its kind in the region as well as medicine valued at millions of Ethiopian birrs. These donations were distributed to various health facilities throughout the Tigray Region to enhance the overall quality of healthcare services (Mekonnen and Lohnert 2018: 11–12).

The ENAHPA is a non-profit organisation founded in November 1999 by Ethiopian-born healthcare professionals residing in the United States and Canada. It focuses on addressing the healthcare needs of women, children, and adults in Ethiopia by providing medical services, promoting preventive health, transfer skills, and offering educational materials and supplies. Every year, ENAHPA sends medical experts to Ethiopia through the MidEth program to conduct workshops, lectures, and training with physicians, nurses, and other medical professionals (Kuschminder and Siegel 2013). ENAHPA also contributes to capacity building and human resource development in the Ethiopian healthcare system, including a significant initiative in 2003 to provide free surgical services to heart disease patients (Degu 2021). The Migration for Development in Ethiopia project, funded by UNDP Ethiopia, seeks to strengthen the Ethiopian government's institutional capacity for accelerating and sustainable development, to eradicate poverty and achieve the Millennium Development, between 2007 and 2011, this project involved 137 health experts from the Ethiopian diaspora educating 262 health professionals in Ethiopia and performing 67 surgical operations in seven hospitals. A donation of 3 million USD in medicines and medical equipment was made, along with 12 short-term assignments at five Ethiopian colleges (IOM 2012).

The Network of Ethiopian Professionals in the Diaspora (NEPID), the North American professional associations that collaborated with the Ethiopian diaspora and medical professional communities. Volunteers carried out a variety of tasks, such as creating guidelines for the nation's medical professionals on how to treat pain, reaching out to international universities, improving and creating medical curricula for Ethiopia's teaching hospitals, and researching the nation's anti-retroviral treatment program. The Ethiopian Diaspora Volunteer Program (EDVP) mobilises diaspora volunteers to build healthcare capacity in their home country, specifically in Ethiopia for HIV/AIDS treatment. Despite being small, the program has successfully placed committed volunteers, offering lessons for similar efforts worldwide. In addition to building schools, diaspora people and organisations also support community centres, health clinics, and schools in rural Ethiopia. For example, TDA, ENAHPA, and AHEAD have made significant contributions to Ethiopia (Mekonnen and Lohnert 2018; Degu 2021).

Ethiopian diaspora investments in pharmaceutical and healthcare production in Ethiopia have been pivotal in addressing healthcare challenges, promoting access to essential medicines, and supporting local manufacturing capacity. The UAE-based pharmaceutical company Gulf Pharmaceutical Industries (Julphar), which formally established a new pharmaceutical production facility in 2013, is one of the noteworthy instances in this field. An additional initiative is Diaspora for Development (ED4D), promoting entrepreneurship among Ethiopian diasporas in the Netherlands or Dutch entrepreneurs collaborating with them. It aims to facilitate skills transfer and short-term return of highly trained individuals to Ethiopia. The initiative involves sending professionals in agro-technology, job development, and fair-trade coffee imports to Ethiopia. The Migration for Development in Ethiopia (MidEth) program, one of the plan's many elements is a program for knowledge and technology transfer. To help Ministries and public institutions, the knowledge and technology transfer effort aims to bring skilled diaspora members back to Ethiopia for a minimum of six months.

TDA Germany makes contributions in the areas of knowledge transfer and experience exchange in addition to delivering medical equipment. 51 physicians and experts were sent by the organisation to the Ayder Referral Hospital (ARH) located in Tigray. TDA has dispatched eight hospital technicians from Germany to maintain the region's X-ray and other medical equipment. TDA Germany also extended invitations to 12 physicians and other clinical staff members from ARH (a total of 15 Ethiopians) to come to Germany for additional training. Today, according to the chairman, ARH is not only the best referral hospital in Tigray but has become one of the best referral hospitals in Ethiopia (Mekonnen and Lohnert 2018: 11–12).

Sahay Solarverein Africa (SSA) is promoting off-grid solar energy in rural Ethiopia by establishing a competence centre at Arba Minch University (AMU) in southern Ethiopia and enabling long-term dissemination of this renewable energy source (Mekonnen and Lohnert 2018: 11). The university collaborates with the University of Applied Sciences and Arts of Southern Switzerland (SUPSI), for training, research and technology transfer in the area of solar energy in solar energy. In 2015, the IOM signed a cooperation agreement with the Ethiopian government to address migration-related challenges and promote economic and social progress. The agreement aims to facilitate the temporary and permanent returns of diaspora professionals' valuable expertise.

The Ethiopian Diaspora Agency is one of the agencies that promotes partnerships between diaspora organisations and local stakeholders and promotes effective philanthropy and social contributions. Professional diasporas from Ethiopia have a great deal of talent and a strong desire to participate in Knowledge, Skill, and Technology Transfer activities. A greater percentage of professional diasporas have the technology, expertise, and knowledge that Ethiopia requires. With virtual, temporary, and permanent return modes of participation, the majority of professional diasporas have expertise, enthusiasm, and plans to help their home nation.

Political Dimension

Diasporic Ethiopians saw themselves as participants in Ethiopia's democratic transition. Ethiopian diasporas in North America, Europe, and other regions of the world have been instrumental in introducing Ethiopian politics in recent decades, both favourably and unfavourably. Ethiopians living abroad have been highly active and have made a conscious effort to spread the values and models of the nations in which they are living. The Ethiopian Proclamation No. 270/2002, was designed to engage Ethiopian expatriates and citizens of Ethiopian origin in the country's development efforts by providing certain rights and privileges. Although it primarily focused on economic and investment opportunities, there is no provision for promoting democratisation in Ethiopia. Empowering the Ethiopian diaspora to economically engage in the country's development, created a pathway for political discourse, advocacy, and the transfer of democratic norms and values.

In addition to sending money home, Ethiopian migrants, who are mostly headed for the US and Europe, attempted to spread democratic and progressive ideals. According to Faist (2008), this is referred to as political remittance. Through their involvement and the sharing of ideas, they demonstrate this political remittance, which in turn contributes to Ethiopia's political progress. Due to their distance from Ethiopian soils, the diasporas play a very limited and dispersed role. Insiders continue to play the most important political role. It appears that the diasporas' primary function is to unwittingly or intentionally promote local activities.

Ethiopian Diasporas have a significant influence on home countries' perceptions and national image due to their extensive media, civic and traditional societies, and financial capacity. In North America, around 80 associations have better financial, mobilisation, and institutional strength. They also operate 82 civic organisations, more than 100 blogs, 300 websites, and 7 audio–video tubes (Brinkerhoff 2019). There are some of them contributing their role in the image building of Ethiopia. A campaign was launched by the Ethiopian diaspora to defend Ethiopia's regional coffee branding against Starbucks' attempts to appropriate these potent cultural icons for financial gain. The effort was led by accomplished diaspora professionals who had extensive networks and a thorough grasp of American industry, government, and culture. They helped organise 100,000 petitions against Starbucks on behalf of Oxfam America and the Ethiopian Intellectual Property Office. Eventually, Starbucks agreed to voluntary trademark licensing arrangements, which were later approved by the US Patent Trade Office (Brinkerhoff 2019). After 2018 Ethiopian government's attitude towards its diaspora shifted towards wider engagement that positive step in increasing the role of Ethiopians abroad in the image building of Ethiopia. The professional diaspora in Ethiopia plays a crucial role in the country's international standing by promoting fair water use and engaging with the global community through various channels, including publications and media. They are encouraged to participate in cultural exchange and image-building through public diplomacy, which has been a part of Ethiopian foreign policy since 2018. Article 5.6 of the Ethiopian diaspora policy advocates the diaspora participating in good governance and democratic transition in a country but the Ethiopian diaspora faces several challenges and barriers when attempting to engage in political activities related to Ethiopia. Legal and regulatory restrictions in host countries can limit their ability to participate in Ethiopian politics, such as citizenship laws that may restrict political involvement or campaigning because Ethiopia does not allow dual citizenship.

Conclusion

Diaspora communities have now become important players in the socio-economic and political paths of their countries of origin in the context of globalisation and transnational migration. The Ethiopian diaspora, estimated at over 2.5 million people, facilitated the attainment of Ethiopia's foreign policies objectives through enabling the country's economic, and social development. Ethiopia's formal policies have sought to channel diaspora engagement into non-political arenas such as investment, education, and skills transfer. Through remittances, investments, and entrepreneurship, the diaspora makes economic contributions that support small companies, reduce poverty, promote economic progress, and households' income. In addition, Ethiopian diaspora have also invested in essential industries such as real estate, manufacturing, and agriculture. The diaspora has also bought bonds to contribute in national infrastructure developments such as the construction of the Grand Ethiopian Renaissance Dam. Aside from the economic contributions, they have also been instrumental as soft power, marketing Ethiopia internationally through cultural exhibitions, business networking forums, and other initiatives. The knowledge transfer and capacity building have been significantly aided by the diaspora. Numerous Ethiopian professionals in foreign countries have collaborated with various institutions and research centres in their home country to further the education and technology development efforts. Ethiopian doctors helped broaden the country's human capital by providing needed services and expertise through numerous programs in the healthcare industry. All these activities demonstrate the wide range of Ethiopian diaspora's contributions to the development of Ethiopia.

References

BenEzer, G. (2002), *The Ethiopian Jewish Exodus: Narratives of the Migration Journey to Israel 1977-1985*, London: Routledge.

Berhanu, K. (2019), "The Political Economy of Diaspora Remittances in the Ethiopian Somali Region", Danish Institute for International Studies Working Paper, Denmark: DIIS.

Bossio, D. A. et al. (2012), "Water implications of foreign direct investment in Ethiopia's agricultural sector", *Water Alternatives*, 5(2), 223–242.

Brinkerhoff, J. M. (2019), "Diasporas and public diplomacy: Distinctions and future prospects", *The Hague Journal of Diplomacy*, 14(1-2): 51–64.

Central Bureau of Statistics (2017), Population of Israel of Ethiopian origin – Collection of data on the occasion of the Sigd Festival: Media release, CBS, Jerusalem. URL: https://www.cbs.gov.il/he/mediarelease/DocLib/2017/334/11_17_334b.pdf [Hebrew]

Chacko, E., & Gebre, P. H. (2013), "Leveraging the diaspora for development: Lessons from Ethiopia", *Geo Journal,* 78: 495–505.

Condé, M.(1998), "Globalization and Diaspora", *Diogenes*, 46(184):29–37.

Degu, T. A. (2021), "Reversing medical brain drain in Ethiopia: thinking beyond restrictive measure", *Mgbakoigba: Journal of African Studies*, 8(2).

EDTF (2024), Ethiopian Diaspora Trust Fund, "Working together to build a better Ethiopia", 20 March, URL: https://www.ethiopiatrustfund.org/.

EIA (2013), Ethiopian Investment Agency, *Invest in Ethiopia: An Investment Guide to Ethiopia—Opportunities and Conditions 2013*, Addis Ababa.

EIC (2019), Ethiopian Investment Commission (2019), Ethiopian Investment Report 2019, Ethiopian Investment Commission, URL: https://eic.waliatechnologies.net/images/Covid-19Response/Covid-19Resources/publications_May-20/EIC-Investment-Report-2019.pdf

Faist, T. (2008), "Migrants as transnational development agents: an inquiry into the newest round of the migration-development nexus", *Population Space and Place*, 14(1): 21–42.

Frankel, J. (1968), *The making of foreign policy: an analysis of decision-making* (Vol. 200). London; New York [etc.]: Oxford UP.

Getahun, S. (2019), "The Role of the Ethiopian Diaspora in Ethiopia", *International Journal of Ethiopian Studies*, 13(2): 215–228.

Goodfellow, T. (2017), "Urban Fortunes and Skeleton cityscapes: Real estate and late urbanization in Kigali and Addis Ababa", *International Journal of Urban and Regional Research*, 41(5): 786–803.

Gulema, S., B. et al. (2024), *The Global Ethiopian Diaspora: Migrations. Connections, and Belongings*, New York: University of Rochester Press.

IOM (2012), Department of Migration Management, "Diaspora engagement projects in the health sector in Africa", IOM International Organisation for Migration.

Kapur, D. (2001), "Diasporas and technology transfer", *Journal of Human Development*, 2(2): 265–286.

Kapur, D. (2003), "Indian Diaspora as a strategic asset", *Economic and Political Weekly*, 38(5): 445–448.

Kapur, D. (2010), *Diaspora development and Democracy: The Domestic Impact of International Migration from India*, New Delhi: Oxford University Press.

Khara, N. K. (2020), "Diaspora and Foreign Policy: A Global Perspective", International Journal of Political Science, 6(4): 2–12.

Kuschminder, K., & Siegel, M. (2013), "Diaspora engagement and policy in Ethiopia" in Michael Collyer (eds.) *Emigration Nations*, London: Palgrave Macmillan.

Mekonnen, M. B., & Lohnert, B. (2018), "Diaspora engagement in development: the case of Ethiopian diaspora associations based in German", *African Diaspora*, 10(1-2): 92–116.

Ministry of Foreign Affairs (2013), Federal Democratic Republic of Ethiopia, *Federal Democratic Republic of Ethiopia Diaspora policy*, Addis Ababa.

Ministry of Information (2020), Federal Democratic Republic of Ethiopia, Foreign Affairs and National Security Policy and Strategy, Addis Ababa.

Ministry of Labor and Social Affairs (2010), Government of Ethiopia, Report on Labor Market Situation and Employment Exchange Service. Addis Ababa.

MPI (2014), Migration Policy Institute, *RAD Diaspora Profile: The Ethiopian Diaspora in the United States,* Washington, DC: Migration Policy Institute.

Ratha, D. et al. (2011), *Leveraging Migration for Africa: Remittances, Skills and Investments,* Washington, DC: World Bank.

Schlenzka, N, (2009), *The Ethiopian diaspora in Germany: its contribution to development in Ethiopia*, Eschborn: Deutsche Gesellschaft fur Technische Zusammenarbeit (GTZ), Economic Development and Employment Division.

Shain, Y. (2007), *Kinship and Diaspora in International Affairs*, Ann Arbor: University of Michigan Press.

Sheffer, G. (1986) "A New Field of Study: Modern Diasporas in International Politics." In G. Sheffer (eds.) *Modern Diasporas in International Politics*, New York: St. Martin's Press.

Shepherd, B. (2013), *The case of Ethiopia: GCC States' Land Investments Abroad. In Summary Report*, Report No. 8, Center for International and Regional Studies.

Solomon, M. (2008), *Determinants of Foreign Direct Investment in Ethiopia*, MA thesis. Maastricht University, Maastricht School of Governance.

Statistics Canada (2017), Census Profile 2016 Census, [Online: web] Accessed on 25 March 2024 URL: https://www12.statcan.gc.ca/census-recensement/2016/dp-pd/prof/index.cfm?Lang=E.

UNCTAD (2023), United Nations Conference on Trade and Development, World Investment Report 2023, [Online: web] Accessed on 18 March 2024 URL: https://unctad.org/data-visualization/global-foreign-direct-investment-flows-over-last-30-years

UNDESA (2017), Trends in International Migrant Stock: The 2017 Revision, United Nations database, United Nations, Department of Economic and Social Affairs, Population Division.
World Bank (2024), Record high remittances to low and middle-income countries in 2023, Washington D.C: The World Bank Group.
Zewde, B. (2002), *A history of modern Ethiopia, 1855–1991*, Addis Ababa: Ohio University Press.

Index

A. Dubey and S. Rajwar (eds.), *Diaspora in Development*, https://doi.org/10.1007/978-981-96-6747-5

FSC
www.fsc.org
MIX
Papier | Fördert
gute Waldnutzung
FSC® C083411

Kolibri 360 GmbH
Ferdinand-Jühlke-Straße 7
99095 Erfurt, Deutschland
produktsicherheit@kolibri360.de